Lecture Notes in Physics

Edited by J. Ehlers, München, K. Hepp, Zürich
R. Kippenhahn, München, H. A. Weidenmüller, Heidelberg
and J. Zittartz, Köln

Managing Editor: W. Beiglböck, Heidelberg

131

Hans C. Fogedby

Theoretical Aspects of Mainly Low Dimensional Magnetic Systems

Springer-Verlag
Berlin Heidelberg New York 1980

Author
Hans C. Fogedby
Institute of Physics, University of Aarhus
DK-8000 Aarhus C, Denmark

ISBN 3-540-10238-8 Springer-Verlag Berlin Heidelberg New York
ISBN 0-387-10238-8 Springer-Verlag New York Heidelberg Berlin

in Germany
d binding: Beltz Offsetdruck, Hemsbach/Bergstr.
-543210

C O N T E N T S

CHAPTER I - INTRODUCTION 1

CHAPTER II - THE ANISOTROPIC MAGNETIC CHAIN IN A LONGITUDINAL FIELD 5

2.1 - General Properties of the Model 5

2.2 - The Energy Spectrum of the Lowest Multiplet for $J^{\perp}=0$ 9

2.3 - The Intensity spectrum of the Lowest Multiplet for $J^{\perp}=0$ 14

2.4 - Concluding Remarks 21

CHAPTER III - THE ISING CHAIN IN A SKEW MAGNETIC FIELD 23

3.1 - General Properties of the Model 23

3.2 - The Energy and Intensity Spectra of the Lowest Multiplet 26

3.3 - Renormalisation Group Analysis of the Energy Spectrum 31

CHAPTER IV - THE HEISENBERG-ISING CHAIN AT ZERO TEMPERATURE 35

4.1 - General Properties of the Model 35

4.2 - The Jordan-Wigner Transformation - The Luttinger Model 37

4.3 - Correlation Functions for the Isotropic XY Model 41

4.4 - The Spectrum of the Isotropic XY Model 50

4.5 - Correlation Functions for the Heisenberg-Ising Chain 52

4.6 - The Spectrum of the Heisenberg-Ising Chain 58

CHAPTER V - HYDRODYNAMICS OF THE HEISENBERG PARAMAGNET 66

5.1 - The Static Description of the Heisenberg Paramagnet 66

5.2 - Conservation Laws - Slow Variables 67

5.3 - Hydrodynamical Equations for the Paramagnet 69

5.4 - Markov Process - Fokker-Planck Equation 71

5.5 - Non-Linear Langevin Equations for the Paramagnet 76

5.6 - Classical Perturbation Theory for the Paramagnet 80

5.7 - Corrections to Hydrodynamics - Long-Time Tails 88

5.8 - Renormalisation Group Treatment 90
5.9 - Concluding Remarks 98

CHAPTER VI - SOLITONS AND MAGNONS IN THE CLASSICAL HEISENBERG CHAIN 100
6.1 - The Model 100
6.2 - Hamiltonian Formulation - Constants of Motion 102
6.3 - Permanent Profile Solution - General Discussion 105
6.4 - Spin Waves - Solitary Waves - Periodic Wave Trains 109
6.5 - The Lax Representation 118
6.6 - The Associated Eigenvalue Problem - General Discussion 120
6.7 - Spectral Properties - Mathematical Discussion 124
6.8 - Scattering States and Bound States 129
6.9 - Time Dependence of the Scattering Data 133
6.10 - The Inverse Scattering Problem 135
6.11 - Canonical Action Angle Variables 138
6.12 - Energy - Momentum - Angular Momentum 146
6.13 - The Spectrum of Solitons and Magnons 149
6.14 - The Infinite Series of Constants of Motion 153
6.15 - Summary and Conclusion 156

LIST OF REFERENCES 158

LIST OF FIGURES

Chapter II page

Fig. 2.1 - The two-cluster state in the second Ising multiplet 6

Fig. 2.2 - The two lowest Ising multiplets (arbitrary units) 6

Fig. 2.3 - The single magnon state, the two magnon bound state, and the two magnon band (arbitrary units) 8

Fig. 2.4 - The energy band at zero field as a function of the total wavenumber (arbitrary units) 11

Fig. 2.5 - Energy versus field for $\nu = 1, 2,$ and 3 (arbitrary units) 13

Fig. 2.6 - The wedge potential and the energy levels of the even manifold (arbitrary units) 14

Fig. 2.7 - The transverse intensity spectrum for $H_o = 0$ (arbitrary units) 18

Fig. 2.8 - The transverse line spectrum for $H_o \gg J^a\cos(ka)$ (arbitrary units) 19

Fig. 2.9 - The transverse line spectrum for $H_o \ll J_a\cos(ka)$ (arbitrary units) 21

Chapter III

Fig. 3.1 - The lowest bands and multiplets of the Ising chain in a transverse and longitudinal field. The dashed lines indicate the maximum extent of the bands for $k = 0$ (arbitrary units) 25

Fig. 3.2 - Energy spectrum of the lowest Ising multiplets showing the transitions induced by the transverse field (arbitrary units) 26

Fig. 3.3 - The energy band for $H^z = 0$ as a function of the total wave number k (arbitrary units) 27

Fig. 3.4 - The lowest Ising multiplet in a skew field (arbitrary units) 28

Fig. 3.5 - The transverse intensity spectrum of the Ising chain in a skew magnetic field (arbitrary units) 31

Fig. 3.6 - Flow of the parameters h and j characterizing the lowest multiplet 34

Chapter IV

Fig. 4.1 - The dispersion law and the half-filled Fermi band characterizing the dynamics of the isotropic XY model 39

Fig. 4.2 - The two-body interaction $-2J'' \cos k$ 40

Fig. 4.3 - The scattering processes contributing to the dynamics of the Heisenberg-Ising chain 40

Fig. 4.4 - The single particle spectrum 42

Fig. 4.5 - The particle-hole spectrum 42

Fig. 4.6 - The multiple particle-hole spectrum 42

Fig. 4.7 - The "light cone" for the isotropic XY model. The asymptotic expressions are valid in the unshaded regions 49

Fig. 4.8 - The longitudinal response of the isotropic XY model as a function of the wavenumber k for fixed frequency ω (arbitrary units) 51

Fig. 4.9 - The transverse response of the isotropic XY model as a function of k for fixed ω (arbitrary units) 52

Fig. 4.10 - The phase velocity $v = \pi((J^{\perp})^2 - (J'')^2)^{\frac{1}{2}} / \mathrm{Arc}\cos(-J''/J^{\perp})$ and the Hartree-Fock approximation $v = 2J^{\perp} - 4J''/\pi$ as a function of $J''/J^{\perp}$ (arbitrary units) 61

Fig. 4.11 - The ferromagnetic longitudinal response of the Heisenberg-Ising chain as a function of the wavenumber k for fixed frequency ω (arbitrary units) 62

Fig. 4.12 - The antiferromagnetic longitudinal response of the Heisenberg-Ising chain as a function of the wavenumber k for fixed frequency ω (arbitrary units) 62

Fig. 4.13 - The static ferromagnetic longitudinal susceptibility for the Heisenberg-Ising chain as a function of the wavenumber k (arbitrary units) 63
Fig. 4.14 - The static antiferromagnetic longitudinal susceptibility for the Heisenberg-Ising chain as a function of the wavenumber k (arbitrary units) 63
Fig. 4.15 - The ferromagnetic transverse response of the Heisenberg-Ising chain as a function of the wavenumber k for fixed frequency ω (arbitrary units) 64
Fig. 4.16 - The static transverse susceptibility for the Heisenberg-Ising chain as a function of the wavenumber k in the case $J^{\perp} > 0$ (arbitrary units) 65

Chapter V
Fig. 5.1 - Markov process, the absence of memory effects is indicated by the dashed line 72
Fig. 5.2 - The conditional transition probability W(x't',x"t") 72
Fig. 5.3 - The exponential decay of a spin fluctuation 82
Fig. 5.4 - The pole structure of $G_0(k\omega)$ in the case of diffusion 82
Fig. 5.5 - The Lorentzian form of the spin correlation function $C_0(k,\omega)$ 83
Fig. 5.6 - The second-order self energy diagram 84
Fig. 5.7 - The second-order noise diagram 86
Fig. 5.8 - The third-order vertex diagrams 87
Fig. 5.9 - The wavenumber shell integration. The shaded area shows the averaged degrees of freedom 92
Fig. 5.10 - The fixed points as a function of the dimension d 94
Fig. 5.11 - The trajectories of $\bar{\lambda}(\ell)$ for different initial values $\bar{\lambda}_0$ for $d > 0$ 95
Fig. 5.12 - The trajectories of $\bar{\lambda}(\ell)$ for different initial values $\bar{\lambda}_0$ for $d < 0$ 95

Chapter VI

Fig. 6.1 - Arbitrary spin configuration with envelope shown 102
Fig. 6.2 - Spin wave motion 106
Fig. 6.3 - The form of F(p) in the case of spin wave motion 107
Fig. 6.4 - Solitary wave motion 107
Fig. 6.5 - The form of F(p) in the case of solitary wave motion 107
Fig. 6.6 - Periodic wave train 108
Fig. 6.7 - The form of F(p) in the case of a periodic wave train 108
Fig. 6.8 - Plot of $\omega = S^z k^2 + h$. The shaded area indicates the spin wave continuum (arbitrary units) 109
Fig. 6.9 - Small amplitude-large width-large velocity solitary wave (arbitrary units) 112
Fig. 6.10 - Large amplitude-small width-small velocity solitary wave (arbitrary units) 112
Fig. 6.11 - The phase ϕ versus $x - vt$ showing the phase shift $\Delta\phi$ (arbitrary units) 112
Fig. 6.12 - The phase shift $\Delta\phi$ versus the phase velocity v (arbitrary units) 112
Fig. 6.13 - The transverse component for a small amplitude solitary wave; the dashed line indicates the envelope (arbitrary units) 113
Fig. 6.14 - The transverse component for a half amplitude solitary wave; the dashed line indicates the envelope (arbitrary units) 113
Fig. 6.15 - The transverse component for a full amplitude solitary wave; the dashed line indicates the envelope (arbitrary units) 113
Fig. 6.16 - The energy density of a solitary wave (arbitrary units) 114
Fig. 6.17 - The velocity momentum relationship for a solitary wave (arbitrary units) 115

Fig. 6.18 - The dispersion law for a solitary wave for $|M^z| = 16, 32, 64$. The shaded area indicates the solitary wave band (arbitrary units) 116

Fig. 6.19 - Periodic wave train solution; the dashed line indicates the envelope (arbitrary units) 117

Fig. 6.20 - Analyticity domain of the Jost functions F and G 123

Fig. 6.21 - Analyticity domain of the transition matrix $T(\lambda)$ 124

Fig. 6.22 - Transmitted and reflected waves for real λ, characterizing a scattering solution 131

Fig. 6.23 - Bound state solution for complex λ 131

Fig. 6.24 - The analytic properties of $a(\lambda)$ 133

Fig. 6.25 - The behaviour of the time dependent phase $Q(\lambda t)$, modulus 2π. 144

Fig. 6.26 - The magnon dispersion law $\omega = \pi^2$ and the soliton dispersion law $E_n = 16\sin^2(\Pi_n/4)/|M_n|$ for two different values of M_n. The shaded area indicates the soliton band (arbitrary units) 152

Fig. 6.27 - Two-soliton collision in configuration space 153

ACKNOWLEDGEMENT

The research reported and summarised in the present "Lecture Series" was initiated at Harvard in 1969, continued in Copenhagen in the early seventies, and completed in Grenoble in the late seventies.

The author wishes to express his gratitude to Professor Paul C. Martin, Harvard University, Professor H. Højgaard Jensen, University of Copenhagen, and Professor Philippe Nozières, Institut Laue-Langevin, Grenoble.

CHAPTER I

INTRODUCTION

The unifying theme of the present extensive paper is theoretical aspects of mainly low dimensional magnetic systems. In the ensuing five chapters we discuss the "anisotropic magnetic chain" (Chapter II), the "Ising chain in a skew field" (Chapter III), the "Heisenberg-Ising chain" (Chapter IV), the "hydrodynamics of the paramagnet" (Chapter V), and "solitons and magnons in the classical Heisenberg chain" (Chapter VI). The main contents of Chapters II to V have already appeared in published form some time ago. These chapters thus essentially constitute a summary. We have, however, included some new material in Chapter III and V. The last chapter deals with more recent work and therefore contains a detailed exposition. With the exception of Chapter VI which is of expository character, we have attempted to omit encumbering details but instead referred to the original papers when necessary. We have, on the other hand, endeavoured to provide enough clues to enable the reader to obtain the main results, so in that sense the chapters are self-contained. We stress that the present paper does not constitute a review of the models cited above but rather an illustration of a variety of important techniques in condensed matter physics.

The emphasis of the paper is on theoretical aspects and only occasionally will experimental information be called upon in order to justify a particular model. The examples considered are, however, all bona fide physical models which in several instances have been applied to the description of actual magnetic materials. With the exception of the paramagnet, the models discussed are all one dimensional. In spite of this limitation the systems do, however, exhibit a rich variety of physical behaviour and require for their understanding the application of several important techniques in quantum and classical mechanics. In our treatment of the anisotropic chain and the Ising chain we thus make use of quantum mechanical secular perturbation theory and quantum mechanical renormalisation group theory. The Heisenberg-Ising chain is a full-fledged many body problem which demands the application of non relativistic quantum field theory. Finally, the classical Heisenberg chain turns out to be an integrable non linear dynamical system which we treat by means of the so-called "inverse scattering method". The hydrodynamics of the paramagnet in arbitrary dimensions is a problem in classical non equilibrium statistical mechanics which calls for the application of classical perturbation theory based on a non linear Langevin description, and which we also treat by means of classical renormalisation group theory. Below we give a brief outline of the content of each chapter.

In Chapter II we consider the anisotropic magnetic chain characterised by the spin half exchange Hamiltonian

$$H = -2J^z \sum_n S_n^z S_{n+1}^z - J^\perp \sum_n (S_n^+ S_{n+1}^- + \mathrm{H.c.}) - J^a \sum_n (S_n^+ S_{n+1}^+ + \mathrm{H.c.}) + H_0 \sum_n S_n^z .$$

For $J^a = 0$ we briefly discuss the single magnon state and derive and solve the integral equation for the two magnon bound state and band for the resulting anisotropic Heisenberg model. For $J^\perp = 0$ and to leading order in J^a/J^z we next apply secular perturbation theory and obtain analytic expressions for the energy levels of the lowest multiplet. In the low field limit the discrete energy spectrum is singular and exhibits a fractional power law dependence on the field H_0 and the effective exchange coupling $J^a \cos(ka)$,

$$E_\nu \propto [J^a \cos(ka)]^{\frac{1}{3}} [H_0]^{\frac{2}{3}} [\nu - \tfrac{1}{4}]^{\frac{2}{3}}, \quad \nu = 1, 2, 3, \ldots$$

By means of a continued fraction techniques we finally derive an analytic expression for the frequency and wavenumber dependent transverse magnetic susceptibility. The relative transverse intensity spectrum is in the low field limit given by $I_\nu \simeq H_0/J^a \cos(ka)$; it is non singular and to leading order independent of the quantum number ν .

The Ising chain in a skew field discussed in Chapter III is governed by the spin half Hamiltonian,

$$H = -2J \sum_n S_n^z S_{n+1}^z - H^z \sum_n S_n^z - \frac{H_x}{2} \sum_n (S_n^+ + S_n^-),$$

which bears a strong resemblance to the one above for the anisotropic chain in the case $J^\perp = 0$. Consequently, to leading order in H^x/J the energy and intensity spectra of the Ising chain have essentially the same analytical form as the corresponding ones for the anisotropic chain, and our discussion here is to some extent a recapitulation of the results obtained in Chapter II. Since the fractional power law behaviour of the energy spectrum for small H^z is indicative of scale invariance we choose, however, the occasion to apply a few modern renormalisation group ideas in order to deduce the form of the spectrum by simple means.

Chapter IV is devoted to a detailed summary of the properties of the Heisenberg-Ising chain characterised by the spin half exchange Hamiltonian,

$$H = -2\sum_n [J^\perp(S_n^x S_{n+1}^x + S_n^y S_{n+1}^y) + J^\parallel S_n^z S_{n+1}^z] ,$$

another variant of the anisotropic chain discussed in Chapter II. The Heisenberg-Ising chain admits, however, a Fermi representation which transform it into an interacting spinless one dimensional fermion gas. Subsequently, the resulting standard many body problem is treated to leading order in $J^\parallel$ by means of a functional integral techniques developed previously for the Tomonaga and Wolff models. We derive the zero temperature transverse and longitudinal correlation functions in the long wavelength-low frequency limit and discuss the spectrum in some detail.

In Chapter V we carry out a detailed analysis of the non linear hydrodynamical corrections to the frequency and wavenumber dependent spin diffusion coefficient $\Gamma(k\omega)$ for the Heisenberg paramagnet in arbitrary dimensions. We first give some background and summarise the theory of stochastic Markov processes. Subsequently, the Fokker Planck and the associated non linear Langevin equations are derived for the coupling of long wavelength-low frequency spin and energy fluctuations in the paramagnet. Neglecting coupling to energy fluctuations, we discuss in some detail the non linear mode coupling equation,

$$\frac{d\bar{S}}{dt} = \lambda \bar{S} \times \nabla^2 \bar{S} + \Gamma \nabla^2 \bar{S} + \bar{\xi} ,$$

using methods developed previously for dynamical critical phenomena. Within the framework of classical perturbation theory pertaining to the above non-linear Langevin equation we compute to second order in the precessional mode coupling strength λ the singular hydrodynamical corrections to the spin diffusion coefficient $\Gamma(k\omega)$. In the long wavelength limit k = 0 and in dimension d we find $\Gamma(k=0,\omega) \simeq \omega^{d/2}$. The spin hydrodynamics of the paramagnet is therefore well-behaved in all physical dimensions, unlike the corresponding calculation for the incompressible Navier Stokes fluid, which yields a correction to the viscosity $\nu(k,\omega)$ diverging in and below two dimensions. With the weak singular hydrodynamical correction is associated a "long time tail" $\Gamma(k=0,t) \simeq 1/t^{1+d/2}$ which in three dimensions behaves as $1/t^{5/2}$, that is different from the famous long time tail $1/t^{3/2}$ for the auto velocity correlations in a fluid. We finally discuss the hydrodynamical corrections within the framework of a classical renormalisation group approach, previously applied to the fluid case.

The final Chapter VI in the present paper deals with the classical isotropic Heisenberg chain in the long wavelength limit, characterised by the continuum Hamiltonian,

$$H = \frac{1}{2}\int \left(\frac{d\bar{S}}{dx}\right)^2 dx,$$

and the resulting non linear precessional evolution equation,

$$\frac{d\bar{S}}{dt} = \bar{S} \times \frac{d^2\bar{S}}{dx^2} \quad .$$

This model is a recent member of the interesting class of completely integrable Hamiltonian systems exhibiting soliton behaviour, which already includes among others the Korteweg-deVries equation, the Sine-Gordon equation, and the non linear Schrödinger equation. We first discuss the permanent profile spin wave, solitary wave, and periodic wave train solutions in some detail. We next give a detailed exposition of the powerful "inverse scattering method" which enables one to essentially diagonalise the system and exhibit a canonical action angle representation. The spectrum is completely exhausted by extended magnon modes, characterised by a quadratic dispersion law $E \propto p^2$, and by localised soliton modes of width $\Gamma \propto 1/E$ and carrying an internal angular momentum m, described by the dispersion law $E \propto \sin^2(p/4)/m$. We also include a derivation of the Gelfand-Levitan-Marchenko equation which in principle allows for the explicit construction of the soliton and magnon modes in configuration space. The chapter is concluded with the derivation of a recursive procedure for the determination of the infinite series of conserved integrated densities.

CHAPTER II

THE ANISOTROPIC MAGNETIC CHAIN IN A LONGITUDINAL FIELD

We begin this chapter by discussing the energy and intensity spectra of the anisotropic magnetic chain within the framework of ordinary secular perturbation theory. As has been shown by Torrance and Tinkham[1-3] the multi magnon bound state spectrum of the linear chain magnetic insulator $CoCl_2 \cdot 2H_2O$ can be interpreted in terms of a one dimensional spin half anisotropic Heisenberg model with a nearest neighbour exchange interaction. Theoretical aspects of this model were subsequently considered by the present author in refs. 4-7. Here we review and summarise the analytical results obtained in refs. 5-7.

2.1 General Properties of the Model

The dynamical properties of the anisotropic magnetic chain are governed by a spin Hamiltonian consisting of three parts,

$$H = H_I + H_H + H_A , \tag{2.1}$$

where

$$H_I = -2J^z \sum_n S_n^z S_{n+1}^z + H_0 \sum_n S_n^z , \tag{2.2a}$$

$$H_H = -J^{\perp} \sum_n (S_n^+ S_{n+1}^- + S_{n+1}^+ S_n^-), \tag{2.2b}$$

$$H_A = -J^a \sum_n (S_n^+ S_{n+1}^+ + S_{n+1}^- S_n^-). \tag{2.2c}$$

With each site n, n=1,..N, of a one dimensional lattice is associated a quantum spin S_n^α, α= x,y,x, $S_n^\pm = S_n^x \pm iS_n^y$, of length 1/2 (in units of $\hbar$) obeying the usual commutator algebra[8] $[S_n^\pm, S_m^z] = \mp S_n^\pm \delta_{nm}$ and $[S_n^-, S_m^+] = -2S_n^z \delta_{nm}$. For the following purposes we need, however, only the non vanishing matrix elements[8] $(S_n^z)_{1/2,1/2} = 1/2$, $(S_n^z)_{-1/2,-1/2} = -1/2$, and $(S_n^+)_{1/2,-1/2} = 1$.

The first term H_I(2.2a) in the Hamiltonian (2.1) is the well-known Ising model[9] in a longitudinal magnetic field H_o (in units such that

$g\mu_B=1$), characterised by the nearest neighbour exchange coupling J^z. The individual spins are constants of motion, S_n^z commutes with H_I, and the dynamics of the model is essentially trivial. The energy spectrum has a simple multiplet structure[1] corresponding to the excitation of clusters of adjacent spin deviations with respect to the aligned ferromagnetic ground state. The lowest multiplet consists of the energy levels $2J^z+nH_0$, n=1,2.., corresponding to single clusters of n spin reversals. We here measure excitation energies relative to the ground state energy. Similarly, the higher multiplets have the energies $2pJ^z+nH_0$, where p is the number of clusters and n the total number of spin deviations. The cluster states encountered in the Ising model can be thought of as a simple kind of dispersionless bound states. In Fig. 2.1 we have shown a state in the second multiplet consisting of two clusters. In Fig. 2.2 we have depicted the two lowest multiplets in a plot of energy versus magnetic field.

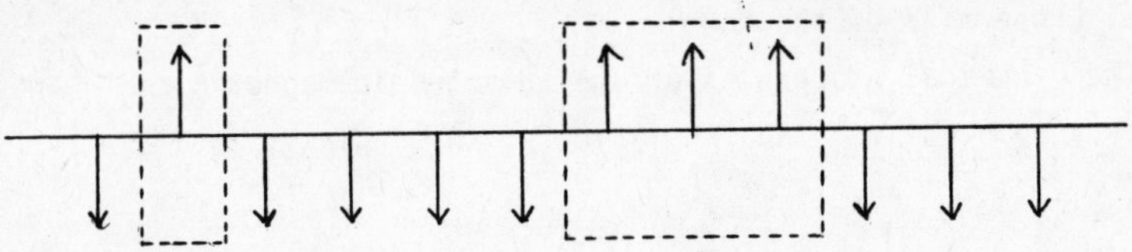

Fig. 2.1 A two cluster state in the second Ising multiplet

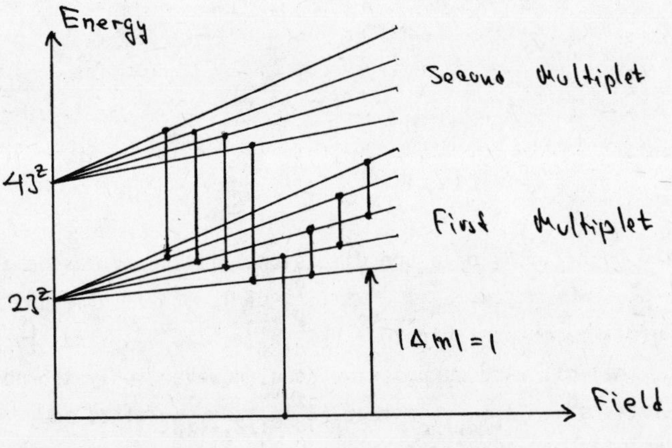

Fig. 2.2 The two lowest Ising multiplets (arbitrary units)

The second terms H_H (2.2b) in the Hamiltonian (2.1) is a transverse mean exchange characterised by the exchange constant $J^{\perp}$. This term is rotationally invariant about the z axis and does therefore not couple the Zeeman split Ising levels. The Heisenberg term H_H, on the other hand, gives rise to dispersion. The first Ising multiplet is transformed into the spectrum of multi magnon bound states for the anisotropic Heisenberg model. In a similar manner the higher multiplets become the multi magnon bands. In the case of the isotropic Heisenberg chan, i.e., $J^z = J^{\perp}$, Bethe[10] found the dispersion law for the general n magnon bound state,

$$E_n(k) = 4J^z \frac{\sin^2(\frac{ka}{2})}{n} + nH_0, \quad n = 1, 2, \ldots \tag{2.3}$$

Here a is the lattice distance (see Fig. 2.1) and the wavenumber k ranges over the first Brillouin zone $-\pi/a \leq k \leq \pi/a$. In the anisotropic case $J^z \neq J^{\perp}$ conventional spin wave theory[11,12] yields the single magnon dispersion law

$$E_1(k) = 2J^z - 2J^{\perp}\cos(ka) + H_0 . \tag{2.4}$$

The energy of the two magnon bound state has been obtained by Orbach[13], see also Wortis[14] and Hanus[15],

$$E_2(k) = 2J^z - 2\frac{(J^{\perp})^2}{J^z}\cos^2(\frac{ka}{2}) + 2H_0 . \tag{2.5}$$

As a prelude to our later considerations it is instructive to derive the dispersion laws (2.4) and (2.5) within the framework of secular perturbation theory[1,8]. Introducing the wave function $\Psi = \sum_n c_n S_n^+ \Psi_0$, where Ψ_0 is the aligned ground state, $\Psi_0^* \Psi_0 = 1$, for a single magnon state and considering the eigenvalue problem $(H_I + H_H)\Psi = E\Psi$ we obtain, using $(S_n^+)_{1/2,-1/2} = 1$, the secular equation,

$$(2J^z + H_0)c_n - J^{\perp}(c_{n+1} + c_{n-1}) = Ec_n . \tag{2.6}$$

The linear difference equation (2.6) is now solved by means of the substitution $c_n \propto \exp(\pm ikan)$ and we arrive at (2.4). The derivation of the dispersion law (2.5) for the two magnon bound state requires a bit of analysis since the two magnon band enters[14]. Introducing the wave function $\Psi = \sum_{nm} c_{nm} S_n^+ S_m^+ \Psi_o$ for the two spin state and considering the eigenvalue problem $(H_I + H_H)\Psi = E\Psi$ we are led to the secular equation

$$(4J^z + 2H_0 - E)C_{nm} - J^\perp(C_{nm+1} + C_{nm-1} + C_{n+1m} + C_{n-1m}) \tag{2.7}$$

$$= 2J^z(\delta_{nm+1} + \delta_{nm-1})C_{nm}, \quad C_{nn} = 0,$$

where the right hand side of (2.7) accounts for the binding of adjacent spin deviations. Subjecting (2.7) to a Fourier transformation in terms of total and relative wavenumbers K and q, and incorporating the constraint $c_{nn} = 0$, arising from the operator identity $(S_n^+) = 0$, by retaining only the odd part in q, we obtain the separable integral equation

$$\left(E_1\left(\frac{K+q}{2}\right) + E_1\left(\frac{K-q}{2}\right) - E\right)C(K,q) = 4J^z \sin\left(\frac{qa}{2}\right)\int_{-2\pi/a}^{2\pi/a} \sin\left(\frac{pa}{2}\right) C(K,p)\frac{dpa}{2\pi} \tag{2.8}$$

The eigenvalue condition yields the two magnon band $E_1((K+q)/2)+E_1((K-q)/2)$ and the two magnon bound state $E_2(K)$. In the anisotropic case $J^\perp < J^z$ the bound state lies below the band for all values of the total wavenumber K. The gap is given by $2(J^z - J^\perp \cos(ka/2))^2/J^z$. In the isotropic case $J^\perp = J^z$ the bound state touches the band at K = 0. In Fig. 2.3 we have shown the single magnon, the two magnon bound state, and the two magnon band in a plot of energy versus total wavenumber for fixed field.

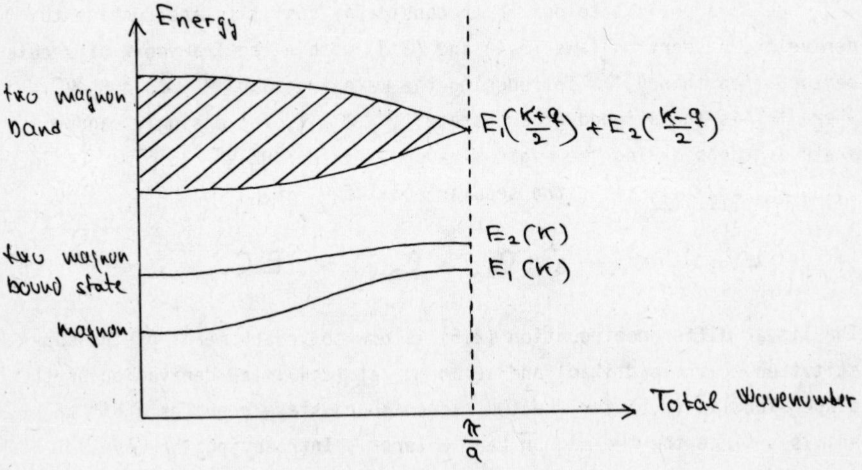

Fig. 2.3 The single magnon state, the two magnon bound state, and the two magnon band (arbitrary units)

An analysis of the three magnon sector requires the solution of a genuine three body problem. The spectrum consists of a three magnon bound state, a band arising from a two magnon bound state and a single magnon, and a band of three single magnons. The problem was considered by the present author[4] to second order in $J^{\perp}/J^{z}$ in the context of a field theoretical approach to the anisotropic chain. The general case has been treated by van Himbergen and Tjon[16], where other references can be found. We finally remark that as shown in ref. 1 the shifts of the levels in the lowest Ising multiplet due to $J^{\perp}$ are for $n \geq 2$ and $J^{\perp} \ll J^{z}$ all of order $(J^{\perp}/J^{z})^2$.

The last term H_A (2.2c) in the Hamiltonian (2.1) is a transverse anisotropy characterised by the exchange constant J^{a}. This term breaks the rotational invariance about the z axis and induces transitions within the Ising multiplets, according to the selection rule $|\Delta m| = 2$ for the magnetic quantum number m. In Fig. 2.2 we have indicated some of the transitions induced by H_A. Owing to the near degeneracy of the levels in the Ising multiplets at low field the transverse anisotropy has a pronounced effect on the spectrum.

The anisotropic magnetic chain is characterised by three dimensionless parameters: $J^{\perp}/J^{z}$, J^{a}/J^{z}, and J^{a}/H_0. $J^{\perp}/J^{z}$ is a measure of the dispersive effects due to H_H, J^{a}/J^{z} characterises the transitions between the lowest multiplet and the ground state caused by H_A, whereas J^{a}/H_0 describes the transitions induced by H_A between the Zeeman split levels of a particular multiplet. In the present chapter we consider only the special case $J^{\perp}/J^{z} = 0$ and, furthermore, limit ourselves to the lowest multiplet assuming $J^{a}/J^{z} \ll 1$ and $H_0/J^{z} \ll 1$. Incidentally, the measurements[1-3] on $CoCl_2 \cdot 2H_2O$ which motivated the research reported in ref. 5-7 are in fact interpreted by choosing both J^{a}/J^{z} and $J^{\perp}/J^{z}$ of order 1/10.

2.2 The Energy Spectrum of the Lowest Multiplet for $J^{\perp} = 0$

The transverse anisotropy H_A has a strong influence on the Ising multiplets, in particular in the low field limit where the levels are nearly degenerate. In order to isolate and analyse this effect we consider the lowest multiplet for $J^{\perp} = 0$. As discussed above the shifts induced by H_H are, with the exception of the lowest magnon which acquires a shift of order $J^{\perp}/J^{z}$, of order $(J^{\perp}/J^{z})^2$ and can be neglected provided $J^{\perp} \ll J^{z}$.

We, furthermore, disregard corrections of order $(J^a/J^z)^2$, i.e., we do not consider transitions between the multiplets and likewise between the lowest multiplet and the ground state.

In a similar manner to the derivation of the single magnon and two magnon bound states in the preceeding section we introduce the wave function $\Psi = \sum_{pn} c_{pn} S_p^+ S_{p+1}^+ \cdot\cdot S_{p+n-1}^+ \Psi_o$ for the nth level in the lowest Ising multiplet. Considering the eigenvalue problem $(H_I + H_A)\Psi = E\Psi$ we obtain, using $(S_p^+)_{1/2,-1/2} = 1$, the secular equation,

$$(2J^z + nH_o)c_{pn} - J^a(c_{pn+2} + c_{pn-2} + c_{p+2n-2} + c_{p-2n+2}) = Ec_{pn}. \tag{2.9}$$

This linear difference equation is partially solved by the substitution $c_{pn} = c_n(k)\exp(\pm ika(p+(n-1)/2))$, i.e.,

$$(2J^z + nH_o)c_n(k) - 2J^a\cos(ka)(c_{n+2}(k) + c_{n-2}(k)) = Ec_n(k), \tag{2.10}$$

where the total wavenumber k runs in the first Brillouin zone $-\pi/a \leq k \leq \pi/a$. The states for even and odd values of n are decoupled (see also Fig. 2.2). In the infinite length limit $N \to \infty$ the expansion coefficients $c_n(k)$ are subject to the boundary conditions $c_0(k) = 0$ and $c_{-1}(k) = 0$ for the even and odd part of the spectrum, respectively.

In the limit of zero field $H_o = 0$ the difference equation (2.1o) is solved by the substitution $c_n(k) \propto \exp(ipa(n-1)/2)$. In the thermodynamic limit we thus obtain the energy band,

$$E(k,p) = 2J^z - 4J^a\cos(ka)\cos(pa), \tag{2.11}$$

where k is the total and p the relative wavenumber of the cluster states. In Fig. 2.4 we have plotted E(k,p) as a function of k. It is interesting to notice that the effective transverse exchange is characterised by $J^a\cos(ka)$ and thus vanishes for $k = \pm\pi/2a$. The maximum value J^a is attained in the long wavelength limit $k = 0$ and at the zone edges $k = \pm\pi/a$. This property is associated with the selection rule $|\Delta m| = 2$ of the transverse anisotropy, i.e., with the symmetry of the exchange Hamiltonian H_A.

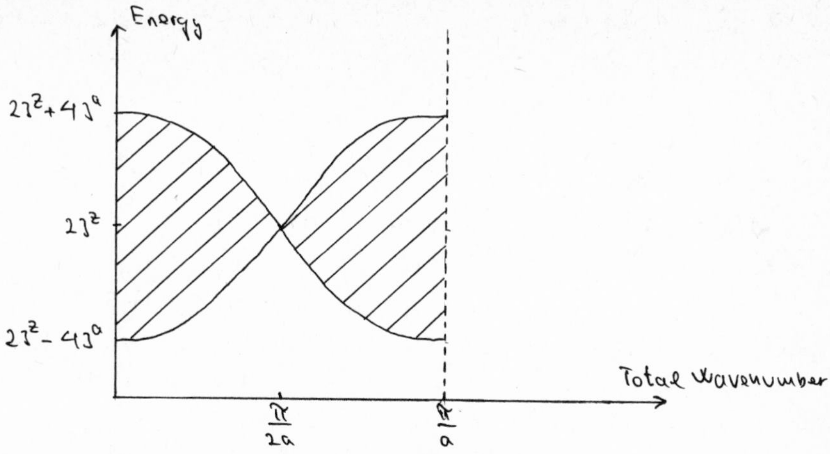

Fig. 2.4 The energy band at zero field as a function of the total wavenumber (arbitrary units)

In the presence of the field H_0 the spectrum determined by (2.10) is discrete and approaches the unperturbed Ising levels $2J^z + nH_0$ for $H_0 \gg J^a$. In the low field limit $H_0 \ll J^a$ the discrete semi-infinite spectrum merges with the bounded energy band E(k,p) Eq. (2.11). In the thermodynamic limit this is a singular transition which shows some interesting features. In ref. 5 it was recognised that equation (2.10) has the same form as a well-known functional recursion formula[17] for the solutions of the Bessel equation[17]. The Bessel function[18] has the correct high field behaviour and we obtain for $J^a\cos(ka) > 0$, $c_n(k) \propto J_{(n-(E-2J^z)/H_0)/2}(2J^a\cos(ka)/H_0)$. For $J^a\cos(ka) < 0$ we have $c_n \rightarrow (-1)^{(n-1)/2}c_n$ for the odd states and $c_n \rightarrow (-1)^{n/2}c_n$ for the even states. Invoking the boundary conditions $c_0(k) = 0$ and $c_{-1}(k) = 0$ the energy levels of the even and odd sector of the multiplet are given by the implicit equations

$$J_{(2J^z-E)/2H_0}(|2J^a\cos(ka)|/H_0) = 0, \quad \text{even sector} \tag{2.12a}$$

and

$$J_{(2J^z-H_0-E)/2H_0}(|2J^a\cos(ka)|/H_0) = 0, \quad \text{odd sector} \tag{2.12b}$$

In the low field limit $H_0 \ll J^a\cos(ka)$ the spectrum is determined by the asymptotic behaviour of the Bessel function for large values of its order and argument[17]. In ref. 5 we obtained to leading order in H_0/J^a,

$$E_\nu = E_{edge} + [18\pi^2]^{\frac{1}{3}} [J^a\cos(ka)]^{\frac{1}{3}} [H_0]^{\frac{2}{3}} [\nu - \tfrac{1}{4}]^{\frac{2}{3}} \qquad (2.13)$$

$$\nu = 1, 2, 3, \ldots$$

where $E_{edge} = 2J^z - 4J^a\cos(ka)$ is the energy of the lower band edge at $H_0 = 0$. The spectrum (2.13) is discrete. The levels are labelled by the quantum number ν which in the intermediate and large field limit, where the expression (2.13) ceases to be valid, becomes identical to the quantum number n of the levels in the unperturbed Ising multiplet. For fixed ν the levels approach the lower band edge E_{edge} asymptotically with infinite slope since $dE_\nu/dH_0 \simeq H_0^{-1/3} \to \infty$ for $H_0 \to 0$. The spectrum exhibits an algebraic singularity at vanishing field of order 2/3 and, furthermore, depends on the effective coupling strength $J^a\cos(ka)$ to the fractional power 1/3. This power law behaviour is non-analytic and cannot be obtained by finite order perturbation theory in J^a.

In the high field limit $H_0 \gg J^a\cos(ka)$ the Zeeman splitting of the Ising levels is large and the correction to the energies can be derived within the framework of ordinary perturbation theory[8]. By means of a well-known series expansion[17] of the Bessel function we obtain in a compact form for the even part of the spectrum

$$\sum_{n=0}^{\infty} \frac{(-1)^n (J^a\cos(ka)/H_0)^{2n}}{\Gamma(n+1)\,\Gamma(n+1+(2J^z-E)/2H_0)} = 0\,. \qquad (2.14)$$

The odd part is given by the substitution $E \to E + H_0$ (see also ref. 7). The expression (2.14) yields the energy spectrum in terms of a power series expansion in $J^a\cos(ka)/H_0$. In Fig. 2.5 we have shown the energy spectrum as a function of H_0.

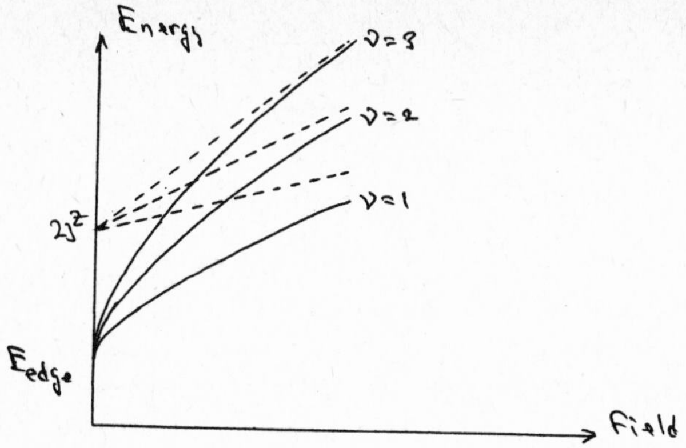

Fig. 2.5 Energy versus field for ν = 1, 2, and 3 (arbitrary units)

There is a simple and instructive way of obtaining the low field spectrum (2.13) without recourse to the theory of Bessel functions using a WKB method[7,19]. Confining ourselves to the even part of the spectrum, measuring E relative to E_{edge}, and denoting $2J^a\cos(ka)$ by J the secular equation (2.10) takes the form $nH_oc_n - J(c_{n+2}+c_{n-2}-2c_n) = Ec_n$, $c_o = 0$. At zero field $H_o = 0$, $E = 4J\sin^2\lambda$ and $c_n \propto \sin(\lambda n)$. In the vicinity of the lower band edge $E \simeq 0$, $\lambda \simeq 0$, and c_n is a slowly varying function of n. Replacing c_n by a continuous function c(n) and expanding $c(n\pm 2)$ about n we obtain for small H_o and small E the one dimensional Schrödinger equation $-4Jc''(n) + nH_oc(n) = Ec(n)$. Incorporating the condition $c(0) = 0$ by introducing an infinite potential wall at $n = 0$, the Schrödinger equation describes the one dimensional motion of a particle with mass 1/8J in the wedge potential[8] $U(n) = nH_o$ for $n > 0$ and $U(n) = \infty$ for $n < 0$. For $H_o > 0$ the motion is bounded and the spectrum is discrete. As H_o approaches zero the motion becomes infinite and the spectrum continuous; this limit is singular. In Fig. 2.6 we have shown the wedge potential U(n).

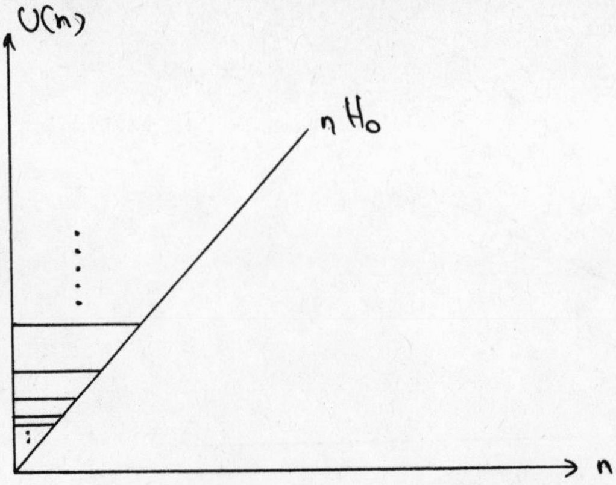

Fig. 2.6 The wedge potential and the energy levels of the even manifold (arbitrary units)

In order to determine the energy levels within the WKB approximation we make use of the Bohr-Sommerfeld quantisation condition[8]

$$\int_0^{E/H_0} p(n)\,dn = \pi(\nu - \tfrac{1}{4}), \quad \nu \text{ large} .$$

Here $p(n) = ((E - nH_0)/4J)^{1/2}$ is the canonical momentum. Integrating and solving for E we obtain the low field spectrum (2.13). Similarly, one can construct the WKB wave function c(n) (see ref. 7).

2.3 The Intensity Spectrum of the Lowest Multiplet for $J^{\perp} = 0$

Owing to the strong admixture effects in the low field limit $H_0 \ll J^a$ the intensity spectrum of the anisotropic chain exhibits some interesting features. The spectrum consists of two parts. For a transverse polarisation of the driving field, corresponding to the selection rule $|\Delta m| = 1$, the odd sector of the manifold is excited directly (see Fig. 2.2). For longitudinal polarisation, corresponding to $|\Delta m| = 0$, the even manifold is indirectly excited owing to its admixture into the aligned ground state. For the purpose of clarity we discuss only the transverse intensity spectrum here. For more details we refer to refs. 6 and 7.

The spin susceptibility $\chi_{nm}^{\alpha\beta}(t,t')$ is defined as the linear response of the spin density $\Delta S_n^{\alpha}(t)$ to an external space and time dependent magnetic field $h_m^{\beta}(t')$ (see, for instance, refs. 20 and 18),

$$\Delta S_n^{\alpha}(t) = \sum_{\beta m} \int_{-\infty}^{\infty} \chi_{nm}^{\alpha\beta}(t,t') h_m^{\beta}(t') dt' .$$

The response is causal and is, in units such that $\hbar = 1$, given by $\chi_{nm}^{\alpha\beta}(tt') = i\langle [S_n^{\alpha}(t), S_m^{\beta}(t')]\rangle \eta(t-t')$, where $\eta(t) = 1$ for $t > 0$ and $= 0$ for $t < 0$, and $\langle .. \rangle$ denotes a ground state expectation value. Using $\int \exp(i\omega t)\eta(t)dt = i/(\omega + i\varepsilon)$, ε positive infinitesimal, we infer the spectral representation

$$\chi_{nm}^{\alpha\beta}(\omega) = \int_{-\infty}^{\infty} \frac{\chi_{nm}^{\alpha\beta''}(\omega')}{\omega' - \omega - i\varepsilon} \frac{d\omega'}{\pi} , \tag{2.15}$$

where

$$\chi_{nm}^{\alpha\beta''}(\omega) = \int_{-\infty}^{\infty} \frac{1}{2} \langle [S_n^{\alpha}(t), S_m^{\beta}(t')]\rangle e^{i\omega(t-t')} d(t-t') \tag{2.16}$$

is the absorptive or dissipative response. The absorbed energy pr. frequency mode is proportional to $\omega\chi_{nm}^{\alpha\beta''}(\omega)$ and the relative intensity spectrum is thus characterised by $\chi_{nm}^{\alpha\beta''}(\omega)$. In general the pole structure of the complex response function $\chi_{nm}^{\alpha\beta}(\omega)$ yields the excitation spectrum and the corresponding residues the relative intensity spectrum.

The transverse intensity spectrum is characterised by

$$\chi_{nm}''(\omega) = \int_{-\infty}^{\infty} \frac{1}{4} \langle [S_n^-(t), S_m^+(t')]\rangle e^{i\omega(t-t')} d(t-t') . \tag{2.17}$$

By means of the Heisenberg equation of motion[8],

$$S_n^-(t) = e^{iHt} S_n^-(0) e^{-iHt} ,$$

we obtain in operator form, absorbing the ground state energy in the Hamiltonian $H = H_I + H_A$,

$$\chi_{nm}''(\omega) = \frac{\pi}{2}\left[\langle S_n^- \delta(\omega - H) S_m^+\rangle - \langle S_m^+ \delta(\omega + H) S_n^-\rangle\right].$$

Neglecting corrections of order $(J^a/J^z)^2$ only the first term contributes, i.e., $\chi''_{nm}(\omega) = \frac{\pi}{2}\Psi_o^* S_n^- \delta(\omega - H) S_m^+ \Psi_o$, where Ψ_o is the aligned ferromagnetic ground state. Inserting $\chi''_{nm}(\omega)$ in the spectral representation (2.15) and noticing that (2.15) allows for an analytic continuation into the upper half plane we have the complex response function $\chi_{nm}(z) = \frac{1}{4}\Psi_o^* S_n^-(H-z)^{-1} S_m^+ \Psi_o$. Fourier transforming, $\chi(kz) = \sum_n \exp(-ika(n-m))\chi_{nm}(z)$, introducing a Bloch basis[1]

$$\Psi_n(k) = \frac{1}{\sqrt{N}}\sum_p e^{ika(p+\frac{n-1}{2})} S_p^+ S_{p+1}^+ \cdots S_{p+n-1}^+ \Psi_o$$

for the Ising multiplet as intermediate states, and expressing $(H-z)^{-1}$ as the quotient of the transposed cofactor $\mathrm{cof}^T(H-z)$ and the determinant $\det(H-z)$, it is easy to show that $\chi(kz)$ in the thermodynamic limit $N \to \infty$ has the form of an infinite continued fraction

$$\chi(k,z) = -\cfrac{\frac{1}{2}}{z - 2J^z - H_o - \cfrac{(2J^a\cos(ka))^2}{z - 2J^z - 3H_o - \cfrac{(2J^a\cos(ka))^2}{z - 2J^z - 5H_o \ddots}}} \qquad (2.18)$$

In the absence of transverse anisotropy $J^a = 0$ or for $k = \pm \pi/2a$ we have the complex response

$$\chi(k,z) = \frac{\frac{1}{2}}{2J^z + H_o - z} \qquad (2.19)$$

i.e.,

$$\chi''(k,\omega) = \frac{\pi}{2}\,\delta(\omega - 2J^z - H_0)\,, \tag{2.20}$$

corresponding to the excitation of the lowest Ising level in accordance with the selection rule $|\Delta m| = 1$. We remark in passing that in the presence of the mean exchange $J^\perp$ this level acquires a dispersion and becomes the single magnon mode with dispersion law $E_1(k)$ (2.4), i.e.,

$$\chi''(k,\omega) = \frac{\pi}{2}\,\delta(\omega - 2J^z + 2J^\perp\cos(ka) + H_0)\,. \tag{2.21}$$

At zero field $H_0 = 0$ the continued fraction (2.18) "repeats itself", i.e.,

$$2\chi(k,z) = \frac{-1}{z - 2J^z + 2(2J^a\cos(ka))^2\,\chi(k,z)}\,.$$

Solving for $\chi(k,z)$ we find

$$\chi(k,z) = \frac{-1}{z - 2J^z + \sqrt{(z-2J^z)^2 - (4J^a\cos(ka))^2}}\,. \tag{2.22}$$

The intensity spectrum is inferred from the spectral representation (2.15),

$$\chi''(k,\omega) = \frac{\sqrt{(4J^a\cos(ka) + 2J^z - \omega)(\omega - 2J^z + 4J^a\cos(ka))}}{(4J^a\cos(ka))^2}\,. \tag{2.23}$$

The zero field intensity spectrum is continouous and has the shape of a semi-ellipse centered about the degeneracy point $2J^z$. The width of the spectrum is $8J^a\cos(ka)$ corresponding to the energy band (2.11) at $H_0 = 0$. The height of the spectrum is $1/4J^a\cos(ka)$. In the limit of vanishing J^a the spectrum shrinks to $\frac{\pi}{2}\delta(\omega - 2J^z)$. In Fig. 2.7 we have shown the intensity spectrum at $H_0 = 0$.

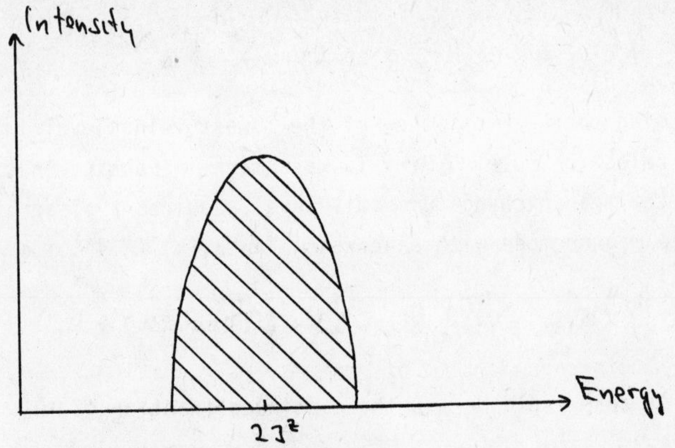

Fig. 2.7 The transverse intensity spectrum for $H_0 = 0$ (arbitrary units)

In the general case of a non-vanishing field the structure of the continued fraction (2.8) implies a non-linear difference equation for $\chi(k,z)$,

$$4(2J^a\cos(ka))^2\,\chi(k,z-2H_0)\,\chi(k,z) + 2(z-2J^z-H_0)\,\chi(k,z) = -1 .$$

The substitution

$$\chi(k,z) = \frac{1}{4J^a\cos(ka)}\,\frac{\Omega(k,z)}{\Omega(k,z+2H_0)}$$

yields a linear equation for $\Omega(k,z)$,

$$\Omega(k,z+2H_0) + \Omega(k,z-2H_0) = \frac{2J^z+H_0-z}{2J^a\cos(ka)}\,\Omega(k,z)$$

of the same form as the difference equation (2.10) discussed in Section 2.2. The correct resonance structure of $\chi(k,z)$ is obtained by choosing the Bessel function solution

$$\Omega(k,z) \propto J_{(2J^z+H_0-z)/2H_0}\left(2J^a\cos(ka)/H_0\right)$$

and we arrive at the expression

$$\chi(k,z) = \frac{1}{4J^a\cos(ka)} \frac{J_{(2J^a+H_0-z)/2H_0}\left(2J^a\cos(ka)/H_0\right)}{J_{(2J^a-H_0-z)/2H_0}\left(2J^a\cos(ka)/H_0\right)} \tag{2.24}$$

for the complex transverse susceptibility.

In the high field limit $H_0 \gg J^a\cos(ka)$ we can terminate the continued fraction (2.18) at a given order in $J^a\cos(ka)/H_0$ and determine the shifts and intensities to that order. Such an approach is, of course, equivalent to the application of ordinary perturbation theory[8]. By means of a series expansion[17] for the Bessel function we obtain

$$\chi(k,z) = \frac{1}{4H_0} \frac{\sum_{n=0}^{\infty} \frac{(-1)^n}{n!} \frac{\left(J^a\cos(ka)/H_0\right)^{2n}}{\Gamma(n+1+(2J^a+H_0-z)/2H_0)}}{\sum_{p=0}^{\infty} \frac{(-1)^p}{p!} \frac{\left(J^a\cos(ka)/H_0\right)^{2p}}{\Gamma(p+(2J^a+H_0-z)/2H_0)}}, \tag{2.25}$$

which expresses $\chi(kz)$ as a ratio of two power series in $J^a\cos(ka)/H_0$. We note that the expansions are convergent for all values of $J^a\cos(ka)/H_0$ and uniformly convergent for all values of z (see ref. 17). In Fig. 2.7 we have depicted the line spectrum in the high field limit.

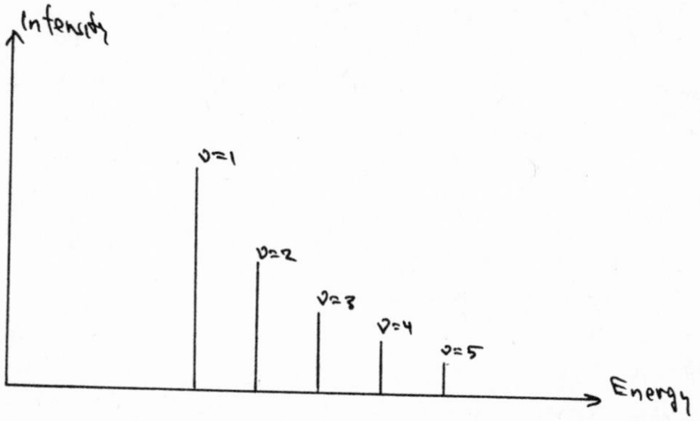

Fig. 2.8 The transverse line spectrum for $H_0 \gg J^a\cos(ka)$ (arbitrary units).

In the low field limit $H_0 \ll J^a \cos(ka)$ the admixture is strong and the total intensity becomes evenly shared among the levels in the multiplet. The analysis of (2.24) is, however, a little technical and we refer to ref. 6 for details. The intensity spectrum is discrete for $H_0 \neq 0$. Expressing $\chi(k,z)$ in the form

$$\chi(k,z) = \sum_{\nu=1}^{\infty} \frac{I_\nu(k)}{E_\nu(k) - z} , \tag{2.26}$$

where $E_\nu(k)$ is given by (2.13),

$$E_\nu(k) = E_{edge} + [18\pi^2]^{\frac{1}{3}} [J^a \cos(ka)]^{\frac{1}{3}} [H_0]^{\frac{2}{3}} [\nu - \tfrac{1}{4}]^{\frac{2}{3}}$$

$$\nu = 1, 2, \ldots$$

and $I_\nu(k)$ is the relative intensity of the νth level,

$$\chi''(k,\omega) = \pi \sum_{\nu=1}^{\infty} I_\nu(k)\, \delta(\omega - E_\nu(k)) \tag{2.27}$$

we find in the low field limit $H_0 \ll J^a \cos(ka)$

$$I_\nu(k) = \frac{H_0}{2J^a \cos(ka)} + O\left(\left(\frac{H_0}{J^a \cos(ka)}\right)^{\frac{4}{3}}\right) \tag{2.28}$$

$$\nu = 1, 2, 3, \ldots$$

We notice that in contrast to the energy spectrum $E_\nu(k)$, the intensity spectrum $I_\nu(k)$ is non-singular in the low field limit. Furthermore, the intensities I_ν are to leading order independent of the quantum number ν indicating that the total intensity is, as anticipated above, evenly shared among the levels. In Fig. 2.8 we have shown the transverse line spectrum in the low field limit.

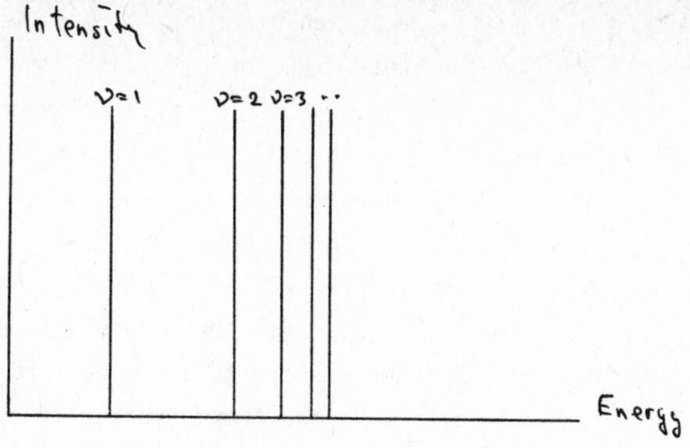

Fig. 2.9 The transverse line spectrum for $H_0 \ll J^a\cos(ka)$ (arbitrary units)

2.4 Concluding Remarks

In the above discussion of the anisotropic magnetic chain we have neglected corrections due to the transverse mean exchange $J^\perp$ and, furthermore, confined our attention to the energy and intensity spectra of the lowest multiplet, thus disregarding transitions between different multiplets, including the ground state singlet, caused by the transverse anisotropy J^a. Whereas the dispersive effects due to $J^\perp$ are small for $J^\perp \ll J^z$ and can be estimated perturbatively[1,4], the transitions between the multiplets, although nominally of order $(J^a/J^z)^2$ for $J^a \ll J^z$, raise some more subtle questions related to the many body character of the system, which we briefly wish to dwell upon.

Owing to the admixture of the lowest multiplet into the ferromagnetic ground state (see Fig. 2.2) the ground state energy is shifted by an amount $\Delta E_0 = -N(J^a)^2/(2J^z+2H_0)$ (refs. 22 and 23) which diverges in the thermodynamic limit $N \to \infty$. This infinite shift has, however, no observable consequences since the lowest multiplet is shifted by a similar amount $\Delta E_n = \Delta E_0 + \Delta_n$, due to its coupling to the second multiplet and to the ground state (see Fig. 2.2). The experimentally accessible excitation energies are thus shifted by Δ_n, n = 1,2,.., which for $J^a \ll J^z$ are small and of order $(J^a/J^z)^2$; the shifts Δ_1, Δ_2, and Δ_3 are given in ref. 4. A similar analysis applies to the other multiplets, they are all shifted by $\Delta E_o + \Delta$, where Δ are small correction of order $(J^a/J^z)^2$ to the transition energies between the multiplets. The infinite shift of the absolute energy scale is a well-known many body effect[24]. In the case of Bose and Fermi systems the infinite shift is absorbed by "normal ordering" the Hamiltonian[25]. The transitions between the multiplets also give rise to divergent wave function corrections proportional to $N^{1/2}$; for the ground state, for instance, $\Delta\Psi_o = -N^{1/2}(J^a/(2J^z+2H_o))\Psi_o$. In the evaluation of the observable intensities I_n, n = 1,2.., for the levels in the first multiple, the N dependent terms, however, cancel to any given order in J^a/J^z and the intensities are well-defined in the thermodynamic limit $N \to \infty$ (ref. 23). The intensities I_1, I_2, and I_3 are given in ref. 4.

CHAPTER III

THE ISING CHAIN IN A SKEW MAGNETIC FIELD

The spin half Ising chain in a skew magnetic field has been treated by the author in ref. 26. The model has a strong resemblance to the anisotropic magnetic chain discussed in the previous chapter. The present review will therefore be rather brief since the results obtained in refs. 5 and 6 can be taken over with essentially a change of variables. We do, however, take the opportunity to discuss the spectrum from the point of view of modern renormalisation group ideas.

3.1 General Properties of the Model

The Ising chain in a skew uniform magnetic field is described by the Hamiltonian

$$H = -2J\sum_n S_n^z S_{n+1}^z - H^z\sum_n S_n^z - H^x\sum_n S_n^x \, , \qquad (3.1)$$

where we without loss of generality have assumed $H^y = 0$. The spin half operators S_n^α, $\alpha = x,y,z$, associated with the lattice point n, n = 1,...N, satisfy the spin algebra[8] $[S_n^\alpha, S_m^\beta] = i\,\delta_{nm}\sum_\gamma \varepsilon^{\alpha\beta\gamma} S_m^\gamma$.

The Ising chain in a longitudinal field H^z,

$$H_I = -2J\sum_n S_n^z S_{n+1}^z - H^z\sum_n S_n^z \qquad (3.2)$$

has already been discussed in Chapter II. The dynamics of the model is trivial and the spectrum has a simple multiplet structure, corresponding to the excitation of clusters of adjacent spin deviations with respect to the aligned ferromagnetic ground state Ψ_0. The energy levels are given by $E_{n,m} = 2nJ + mH^z$, where n = 1,2,.. is the number of spin clusters and m = 1,2,.. the total number of spin reversals (see Figs. 2.1 and 2.2).

The Ising chain in a transverse field H^x

$$H_T = -2J\sum_n S_n^z S_{n+1}^z - H^x \sum_n S_n^x \tag{3.3}$$

on the other hand, is a non-trivial quantum model [27-29]. Pikin and Tsukernik[27] and Pfeuty[28] have shown that the Hamiltonian (3.3) has the same spectrum as a non-interacting Fermi gas

$$H_F = \sum_k E_k (a_k^\dagger a_k - \tfrac{1}{2}) \tag{3.4}$$

with the single particle dispersion law

$$E_k = \sqrt{J^2 + (H^x)^2 - 2JH^x \cos(ka)} \,. \tag{3.5}$$

Here a_k and $a_k^\dagger$ are fermion operators in wavenumber space $|k| \leq \pi/a$, a being the lattice parameter, satisfying the anti-commutation relations $\{a_k, a_p^\dagger\} = \delta_{kp}$ and $\{a_k, a_p\} = 0$. The correspondence is accomplished by means of a Jordan-Wigner transformation[30], which we discuss in Chapter IV, followed by a Bogoliubov-Valatin transformation[31]. For a closed chain with cyclic boundary conditions the spectrum is composed of bands, corresponding to the excitation of an even number of fermions. We remark, however, that although the spectrum of (3.3) has a simple structure, the corresponding wave functions are complicated, owing to the non linear character of the Jordan-Wigner transformation, and the evaluation of, for instance, the transverse spin correlation function pertaining to (3.3) is non trivial. In Fig.3.1 we have shown the two lowest bands of the Ising chain in a transverse field. In the same figure we have also depicted the two lowest multiplets of the Ising chain in a longitudinal field.

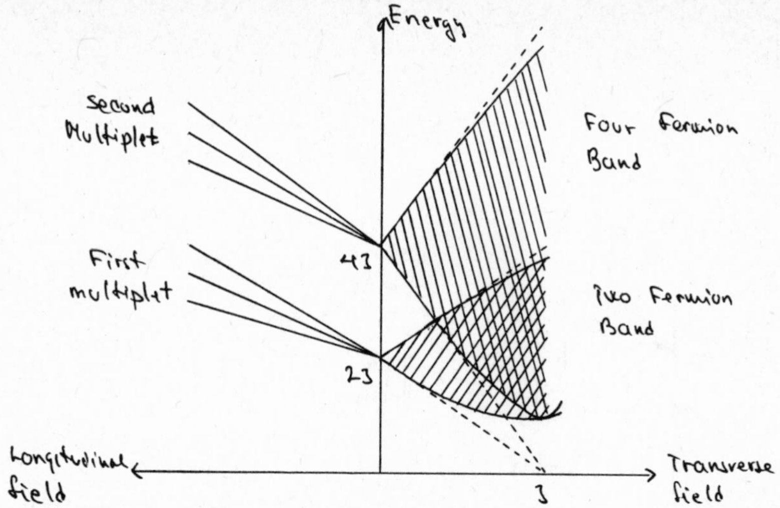

Fig. 3.1 The lowest bands and multiplets of the Ising chain in a transverse and longitudinal field. The dashed lines indicate the maximum extent of the bands for k = 0 (arbitrary units)

For H^x less that the critical field value J the Ising term $-2J\sum_n S_n^z S_{n+1}^z$ dominates and the model displays long range order in the ground state. For H^x = J and in the long wavelength limit k = 0 the fermion bands hybridise with the ground state (see Fig. 3.1) and the model undergoes a zero temperature phase transition to a state with no long range order, i.e., the Zeeman term $-H^x\sum_n S_n^x$ dominates the Ising term. The critical exponents associated with this transition have recently been evaluated within the framework of a real space renormalisation group approach[32].

In the general case of a skew field the model (3.1) is quite similar to the anisotropic magnetic chain

$$H' = -2J^z\sum_n S_n^z S_{n+1}^z + H_0\sum_n S_n^z - J^a\sum_n (S_n^+ S_{n+1}^+ + S_n^- S_{n+1}^-),$$

discussed in the previous chapter. The transverse field term

$$-H^x\sum_n S_n^x = \frac{1}{2}H^x\sum_n (S_n^+ + S_n^-), \qquad (3.6)$$

plays the same role as the transverse anisotropy J^a; it induces transitions between the energy levels of the Ising multiplets, here however, according to the selection rule $|\Delta m| = 1$. In Fig.3.2 we have shown the two lowest Ising multiplets and some of the transitions induced by the transverse field.

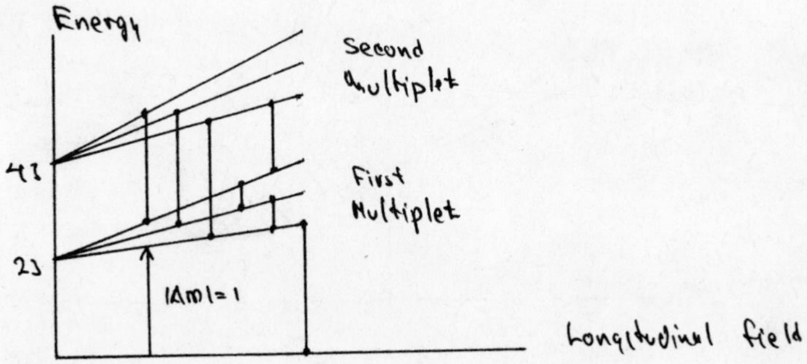

Fig.3.2 Energy spectrum of the lowest Ising multiplets showing the transitions induced by the transverse field (arbitrary units)

The dynamics of the model is determined by the parameters H^x/J and H^z/J, characterising the strength of the field compared to the exchange coupling, and by the parameter H^x/H^z, depending on the direction of the field. In contrast to H^x/J and H^z/J which are bounded and can be treated perturbatively, the ratio H^x/H^z is unrestricted and assumes large values in the limit of a vanishing longitudinal field. In close analogy with our treatment of the anisotropic chain in Chapter II we consider here only the weak field case $H^x, H^z \ll J$ and neglect corrections of order $(H^x/J)^2$ and $(H^z/J)^2$, i.e., we disregard transitions and hybridisations between neighbouring multiplets, including the ground state singlet. As in the case of the anisotropic chain the interesting physics of the model is determined by the ratio H^x/H^z which we treat to "all orders" by providing analytical expressions for the energy and intensity spectra pertaining to the lowest multiplet.

3.2 The Energy and Intensity Spectra of the Lowest Multiplet

The transverse field term (3.6) induces transitions between the Zeeman split Ising levels of $H_I(3.2)$. Introducing the wave function

$$\Psi = \sum_{pn} c_{pn} S_p^- S_{p+1}^- \cdots S_{p+n-1}^- \Psi_0$$

for the states in the lowest Ising multiplet and considering the eigenvalue problem $H\Psi = E\Psi$, using $(S_p^+)_{1/2,-1/2} = 1$, we obtain the secular equation

$$(2J + nH^z)c_{pn} - \frac{1}{2}H^x(c_{pn+1} + c_{pn-1} + c_{p+1\,n-1} + c_{p+1\,n+1}) = E\,c_{pn} \tag{3.7}$$

By means of the substitution $c_{pn} = c_n(k)\exp(ika(p+(n-1)/2))$ we are led to

$$(2J+nH^z)c_n(k) - H^x\cos(ka/2)(c_{n+1}(k) + c_{n-1}(k)) = E\,c_n(k) \tag{3.8}$$

At zero longitudinal field $H^z = 0$ Eq.(3.8) is solved by the substitution $c_n(k) \propto \exp(ipa(n-1)/2)$ yielding the energy band

$$E(k,p) = 2J - 2H^x\cos(ka/2)\cos(pa/2), \tag{3.9}$$

a result which is in agreement with the two fermion band $E_{k'} + E_{p'}$ for $H^x \ll J$ (see Fig.3.1). By comparison we infer that $k = p' + k'$ is the total and $p = p' - k'$ the relative wave numbers of the fermion pair. Invoking the boundary condition $c_0(k) = 0$ the expansion coefficient for the band is $c_n(k) \propto \sin(pan/2)$ and we obtain the wave function

$$\Psi_{band} \propto \sum_{np} \sin\left(\frac{pan}{2}\right) e^{ika(p+\frac{n-1}{2})} S_p^- S_{p+1}^- \cdots S_{p+n-1}^- \Psi_0. \tag{3.10}$$

We notice that the effective transverse field is given by $H^x\cos(ka/2)$ and thus vanishes at the zone edges $k = \pm\pi/a$. In Fig.3.3 we have shown the energy band for $H^z = 0$ as a function of the total wave number k.

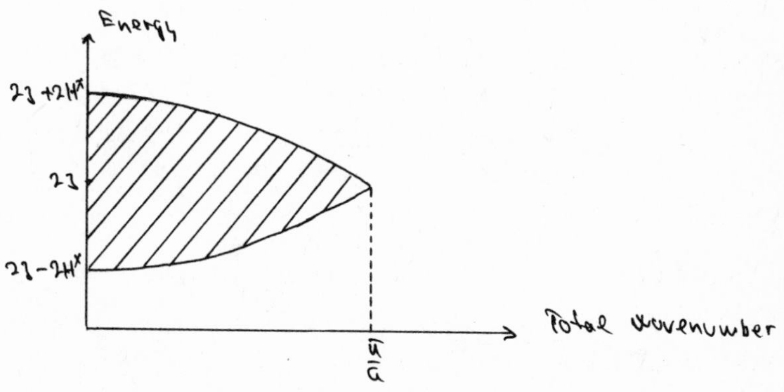

Fig.3.3 The energy band for $H^z = 0$ as a function of the total wave number k (arbitrary units).

In the general case the secular equation (3.8) has the same form as the eigenvalue equation (2.10) discussed in Chapter II, and we immėdiately infer the solution (in the thermodynamic limit).

$$c_n(k) \propto J_{n-(E-2J)/H^z}\left(2H^x\cos(ka/2)/H^z\right). \tag{3.11}$$

The energy spectrum follows from the boundary condition $c_0(k) = 0$ and is given by the implicit equation

$$J_{(2J-E)/H^z}\left(2H^x\cos(ka/2)/H^z\right) = 0. \tag{3.12}$$

In the limit of a weak longitudinal field $H^z \ll H^x\cos(ka/2)$ the spectrum exhibits algebraic singularities as a function of H^z and $H^x\cos(ka/2)$,

$$E_\nu(k) = 2J - 2H^x\cos\left(\frac{ka}{2}\right) + \left[\frac{3\pi}{2}\right]^{\frac{2}{3}}\left[H^x\cos\left(\frac{ka}{2}\right)\right]^{\frac{1}{3}}\left[H^z\right]^{\frac{2}{3}}\left[\nu-\frac{1}{4}\right]^{\frac{2}{3}}$$
$$\nu = 1, 2, \ldots \tag{3.13}$$

In Fig.3.4 we have, in a three-dimensional plot, shown the lowest multiplet of the Ising chain in a skew magnetic field.

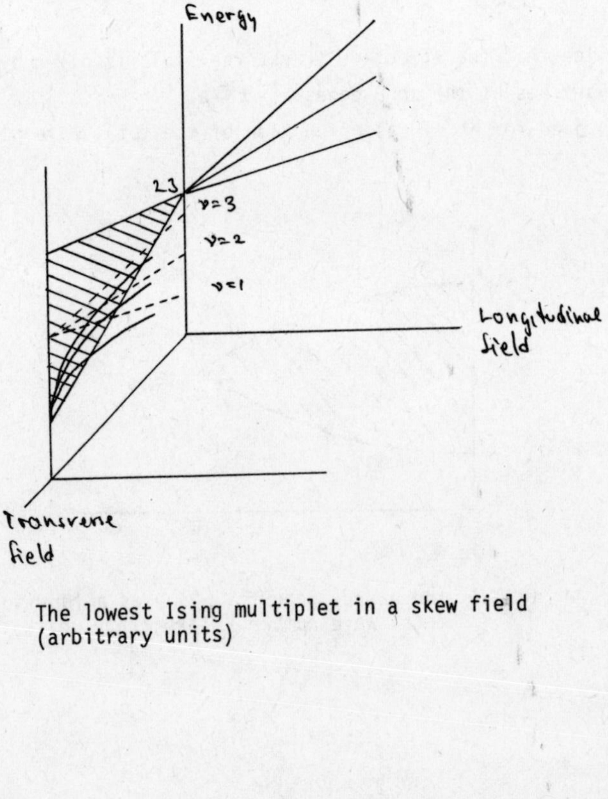

Fig.3.4 The lowest Ising multiplet in a skew field (arbitrary units)

The transverse complex susceptibility is defined as[20,21]

$$\chi(k,z) = \int_{-\infty}^{\infty} \frac{\chi''(k,\omega)}{\omega - z}\,\frac{d\omega}{\pi}, \tag{3.14}$$

where

$$\chi''(k,\omega) = \int_{-\infty}^{\infty} \frac{1}{4}\langle [S_n^-(t), S_m^+(t')]\rangle e^{i\omega(t-t')}\, d(t-t') \tag{3.15}$$

is the absorptive response. Following the procedure in Chapter II, it is easy to show that $\chi(k,z)$ has the form of a continued fraction,

$$\chi(k,z) = \cfrac{-\frac{1}{2}}{z - 2J - H^z - \cfrac{(H^x\cos(ka/2))^2}{z - 2J - 2H^z - \cfrac{(H^x\cos(ka/2))^2}{z - 2J - 3H^z \ldots}}} \tag{3.16}$$

and as a result satisfies the non-linear difference equation

$$4(H^x\cos(ka/2))^2\chi(k,z)\chi(k,z-H^z) + 2(z-2J-H^z)\chi(k,z) + 1 = 0 \tag{3.17}$$

For vanishing transverse field $H^x = 0$, or at the zone edges $k = \pm\pi/a$,

$$\chi(k,z) = \frac{1}{2}\,\frac{1}{2J + H^z - z}, \tag{3.18}$$

and the absorptive response is given by

$$\chi''(k,\omega) = \frac{\pi}{2}\,\delta(\omega - 2J - H^z), \tag{3.19}$$

corresponding to the excitation of the lowest Ising level, according to the selection rule $|\Delta m| = 1$. In the case of zero longitudinal field $H^z = 0$ the expression (3.17) reduces to a quadratic equation for $\chi(k,z)$ with the solution

$$\chi(k,z) = -\frac{1}{z - 2J + \sqrt{(z-2J)^2 - (2H^x\cos(ka/2))^2}} \tag{3.20}$$

and the spectral representation (3.14) implies the absorptive response

$$\chi''(k,\omega) = \frac{\sqrt{(2H^x\cos(ka/2) + 2J - \omega)(\omega - 2J + 2H^x\cos(ka/2))}}{(2H^x\cos(ka/2))^2}. \tag{3.21}$$

The intensity spectrum is continuous and has the shape of a semi-ellipse of width $4H^x\cos(ka/2)$ and height $(2H^x\cos(ka/2))^{-1}$ centered about the degeneracy point 2J.

In the general case of a skew field the difference equation (3.17) is solved by the substitution

$$\chi(k,z) = \frac{1}{2H^x\cos(ka/2)}\,\frac{\Omega(k,z)}{\Omega(k,z+H^z)},$$

where $\Omega(kz)$ in turn satisfies

$$\Omega(k,z+H^z)+\Omega(k,z-H^z) = \frac{2J+H^z-z}{H^x\cos(ka/2)}\,\Omega(k,z),$$

i.e., the functional recursion relation for the solutions to the Bessel equation[33]. In the thermodynamic limit $N\to\infty$ we thus obtain the complex transverse susceptibility

$$\chi(k,z) = \frac{1}{2H^x\cos(ka/2)}\,\frac{J_{(2J+H^z-z)/H^z}\left(2H^x\cos(ka/2)/H^z\right)}{J_{(2J-z)/H^z}\left(2H^x\cos(ka/2)/H^z\right)}. \tag{3.22}$$

The intensity spectrum is discrete. In the limit of a small longitudinal field $H^z \ll H^x\cos(ka/2)$

$$\chi(k,z) = \sum_{\nu=1}^{\infty}\frac{I_\nu(k)}{E_\nu(k)-z}, \tag{3.23}$$

where $E_\nu(k)$ is given by Eq.(3.13), and $I_\nu(k)$ is the relative intensity of the νth level,

$$\chi''(k,\omega) = \pi\sum_{\nu=1}^{\infty} I_\nu(k)\,\delta(\omega-E_\nu(k)). \tag{3.24}$$

To leading order we find

$$I_\nu(k) = \frac{H^z}{2H^x\cos(ka/2)} + O\left(\left(\frac{H^z}{H^x\cos(ka/2)}\right)^{\frac{4}{3}}\right) \tag{3.25}$$

$\nu = 1, 2, \cdots$

The line spectrum I_ν is non singular in the limit of vanishing longitudinal field and, furthermore, to leading order independent of the quantum number ν, i.e., the total intensity becomes evenly shared among the near degenerate levels. In Fig.3.5 we have, in a three dimensional plot, summarised the properties of the transverse intensity spectrum.

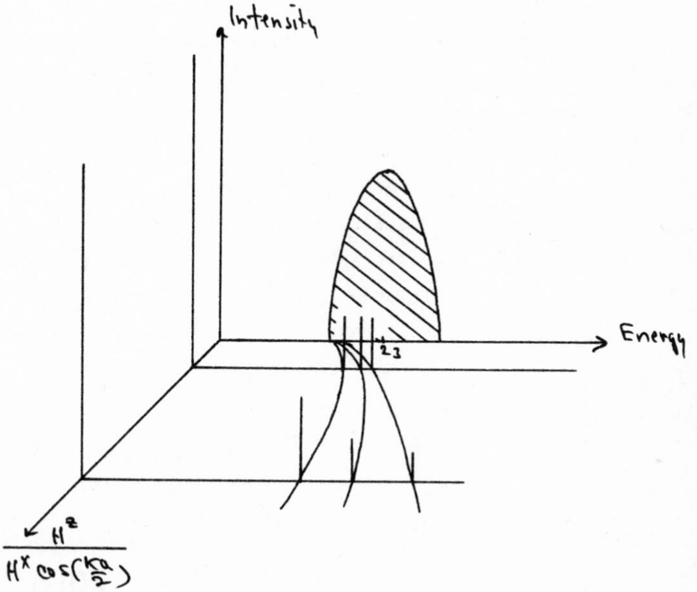

Fig.3.5 The transverse intensity spectrum of the Ising chain in a skew magnetic field (arbitrary units)

3.3 Renormalisation Group Analysis of the Energy Spectrum

It is instructive to discuss the energy spectrum and to derive the characteristic power law behaviour in an alternative manner employing a few modern renormalisation group ideas[34-36].

Introducing the notation $2J = \Delta$, $H^x\cos(ka/2) = j$, and $H^z = h$, the secular equation (3.8) takes the form

$$\Delta c_n + nhc_n - j(c_{n+1} + c_{n-1}) = Ec_n . \tag{3.26}$$

Since the energy levels are equidistant for $j = 0$ we can associate an effective non-linear harmonic oscillator Hamiltonian H with the eigenvalue problem (3.26). Introducing a Bose operator b, $[b,b^+] = 1$, and noting that $\langle n-1|b|n\rangle = 1$, we have

$$H = \Delta + h b^{\dagger} b - j\left(\frac{1}{\sqrt{1+b^{\dagger}b}}\, b + b^{\dagger}\,\frac{1}{\sqrt{1+b^{\dagger}b}}\right), \qquad (3.27)$$

where $(1+b^{\dagger}b)^{-1/2}$ is defined by its power series expansion. The non-linear Bose Hamiltonian (3.27) is characterised by three parameters: The shift Δ, the field h (the longitudinal field), and the coupling j (the transverse field), all of dimension energy, i.e., $\dim(\Delta) = \dim(h) = \dim(j) = \dim(E)$.

A renormalisation group analysis of a quantum mechanical eigenvalue problem $H\Psi = E\Psi$ exhibiting scale invariance, i.e., power law behaviour, proceeds in three steps. Firstly, one eliminates a set of states from that part of the spectrum which does not show scale invariance. Secondly, one scales the problem and introduces approximations such that the Hamiltonian for the new "diluted" and scaled eigenvalue equation $H'\Psi' = E\Psi'$ has the same form as the original Hamiltonian H. The third step consists in iterating the above procedure and establishing renormalisation group equations for the parameters in the Hamiltonian. Notice that this program is completely analogous to the elimination of short wavelength degrees of freedom in the renormalisation group theory of critical phenomena[36], (see also our treatment of hydrodynamical corrections in Chapter V).

Let us now treat the secular equation (3.26) from the above point of view by first eliminating every second level. From (3.26) we have

$$(\Delta + (n \pm 1)h - E)c_{n \pm 1} = j(c_{n \pm 2} + c_n).$$

Solving for $c_{n \pm 1}$ and inserting in (3.26) we obtain the more complicated eigenvalue problem

$$\left(\Delta + nh - E - \frac{j^2}{\Delta + (n+1)h - E} - \frac{j^2}{\Delta + (n-1)h - E}\right)c_n =$$

$$\frac{j^2}{\Delta + (n+1)h - E}\, c_{n+2} + \frac{j^2}{\Delta + (n-1)h - E}\, c_{n-2}.$$

Notice that we have so far not made any approximations. The appearence of energy-dependent denominators shows that the elimination of states in general gives rise to retardation effects. In order to avoid complications related to the boundary condition $c_0 = 0$ we now make the first approximation by assuming $n \gg 1$, i.e., we confine our analysis to the interior of the spectrum. The eigenvalue equation then reduces to

$$[(\Delta + nh - E)^2 - 2j^2]c_n = j^2[c_{n+2} + c_{n-2}]$$

In order to compare the "diluted" eigenvalue problem above to the original one (3.26) one scales the spectrum by introducing the new expansion coefficient $c'_n = c_{2n}$, i.e.,

$$[(\Delta+2nh-E)^2-2j^2]c'_n = j^2[c'_{n+1}+c'_{n-1}]$$

or,

$$[(\Delta-E)^2-2j^2+4nh(\Delta-E)+4n^2h^2]c'_n = j^2[c'_{n+1}+c'_{n-1}].$$

In the low field limit $nh \ll j$ we obtain, neglecting corrections of order $(nh/j)^2$, a renormalised eigenvalue problem

$$\left[\left(\frac{\Delta}{j}-\frac{E}{j}\right)^2-2+4n\frac{h}{j}\left(\frac{\Delta}{j}-\frac{E}{j}\right)\right]c'_n = c'_{n+1}+c'_{n-1} \qquad (3.27)$$

of the same form as Eq.(3.26),

$$\left[\frac{\Delta}{j}-\frac{E}{j}+n\frac{h}{j}\right]c_n = c_{n+1}+c_{n-1}.$$

Comparing the two equations and denoting the renormalised parameters by Δ', h', and j' we finally infer the renormalisation group equations

$$\frac{\Delta'}{j'}-\frac{E}{j'} = \left(\frac{\Delta}{j}-\frac{E}{j}\right)^2-2 \qquad (3.28a)$$

$$\frac{h'}{j'} = 4\frac{h}{j}\left(\frac{\Delta}{j}-\frac{E}{j}\right). \qquad (3.29b)$$

One notices that the renormalised parameters Δ', j', and h' in general depend on the energy E, indicating the presence of retardation effects.

Let us now examine the renormalisation-group equations (3.28a) and (3.28b). Since the Hamiltonians $H = H(\Delta,h,j)$ and $H' = H(\Delta',h',j')$, with the above assumptions $n \gg 1$ and $nh \ll j$, describe the same physical spectrum we can iterate the equations (3.28a) and (3.28b), i.e., Δ,h,j $\Delta',h',j' \to \Delta'',h'',j''$ .. , or, equivalently, $H \to H' \to H''$ ... The interesting case of scale invariance arises when the iteration leads to a fixed point Hamiltonian $H^* = H(\Delta^*,h^*,j^*)$. Since we have only two equations for three parameters we chose the trajectory of Δ in parameter space such that $\Delta = 2j$ and $\Delta' = 2j'$. This corresponds to fixing the lower band edge at $E = 0$, and is completely equivalent to choosing $T = T_c$ in the analogous treatment of critical phenomena[36]. With this choice of Δ we obtain the following renormalisation group equations for j and h:

$$j' = \frac{j^2}{4j - E}, \tag{3.29a}$$

$$h' = 4h\,\frac{2j - E}{4j - E}. \tag{3.29b}$$

Let us investigate the bottom of the spectrum in the vicinity of the lower band edge E = 0. For $E \ll j$ we obtain the equations $j' = \frac{1}{4}j$ and $h' = 2h$, which by iteration have an unstable fixed point at $(h^*, j^*) = (0,0)$. The physical properties, in particular, the energy spectrum, must be scale invariant under the renormalisation group. From the equations $j' = \frac{1}{4}j$ and $h' = 2h$ we infer that $j'h'^2 = jh^2$, i.e., jh^2 is a scale invariant, and the discrete energy spectrum E_ν is, consequently, a function of jh^2. Since $\dim(E) = \dim(j) = \dim(h)$ we obtain the power law behaviour

$$E_\nu \simeq [jh^2]^{\frac{1}{3}} = [j]^{\frac{1}{3}}[h]^{\frac{2}{3}}$$

in agreement with (3.13). In the high field limit $h \gg j$ the spectrum $E_n = nh$ is trivially scale invariant, i.e., $E \propto h$, and we infer the stable fixed point $(h^*, j^*) = (\infty, 0)$.

The above renormalisation group analysis is restricted to the low field-large n limit, $nh \ll j$ and $n \gg 1$. At intermediate fields the renormalisation group equations become more complicated due to retardation effects, and one must in general enlarge the number of non-linear terms in the corresponding Hamiltonian. The Hamiltonian flow[35] leaves the two-parameter space (j,h) and enters a multi-dimensional region in parameter space. For large fields $h \gg j$ the trajectories "return" to the two-dimensional parameter space and approach the stable fixed point $(h^*, j^*) = (\infty, 0)$. In Fig.3 we have shown the flow of the parameters h and j.

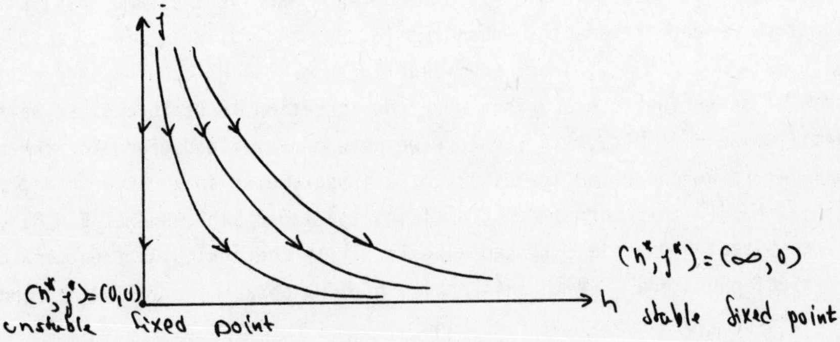

Fig.3.6 Flow of the parameters h and j characterising the lowest multiplet

CHAPTER IV

THE HEISENBERG-ISING CHAIN AT ZERO TEMPERATURE

In the preceeding two chapters we discussed the "anisotropic magnetic chain" and the "Ising chain in a skew field", using ordinary time-independent perturbation theory and a little renormalisation group theory. In the present chapter we turn to a field theoretical investigation of the one-dimensional Heisenberg-Ising model, another variant of the magnetic chain. This model has been treated by the author in ref.37 by means of a functional-integral technique developed previously for the Tomonaga and Wolff models in refs. 38 - 40.

4.1 General Properties of the Model

The dynamical properties of the Heisenberg-Ising chain is characterised by the exchange Hamiltonian

$$H = -2\sum_n [J^{\perp}(S_n^x S_{n+1}^x + S_n^y S_{n+1}^y) + J^{\parallel} S_n^z S_{n+1}^z], \qquad (4.1)$$

where S_n^{α}, $\alpha = x,y,z$, is a spin half field[8], $[S_n^{\alpha}, S_p^{\beta}] = i\delta_{np}\sum_{\gamma}\varepsilon^{\alpha\beta\gamma}S_n^{\gamma}$, associated with the site n of a one-dimensional lattice, n = 1,..N. The model (4.1) is the special case $H_0 = J^a = 0$ of the anisotropic magnetic chain (2.1) discussed in Chapter II. In the present context we consider, however, the opposite limit $J^z = J^{\parallel} \ll J^{\perp}$.

For vanishing longitudinal exchange $J^{\parallel} = 0$ the Hamiltonian (4.1) reduces to the isotropic limit $\alpha = 1$ of the general XY model,

$$H_{XY} = -2\sum_n J^{\perp}(S_n^x S_{n+1}^x + \alpha S_n^y S_{n+1}^y) \qquad (4.2)$$

which was introduced by Lieb et al.[31] for the purpose of investigating the influence of anisotropy on the ground state correlations. For $\alpha = 0$ we obtain the Ising model[9] $H = -2J\sum_n S_n^x S_{n+1}^x$ which exhibits long range order in the ground state. In the isotropic limit, $\alpha = 1$, it was shown in ref.31 that the static longitudinal correlations in the ground state Ψ_0 decay algebracially,

$$\Psi_0^* S_n^z S_m^z \Psi_0 \simeq \frac{(-1)^{n-m} - 1}{|n-m|^2} .$$

Subsequently, McCoy[41] evaluated the static longitudinal and transverse spin correlations in the general anisotropic case at finite temperature. For the ground state correlations, in particular, he finds, for $\alpha < 1$,

$$\Psi_0^* S_n^z S_m^z \Psi_0 \simeq \frac{\alpha^{|n-m|}}{|n-m|^2} \quad \text{for } |n-m| \to \infty$$

$$\Psi_0^* S_n^x S_m^x \Psi_0 \simeq \sqrt{1-\alpha^2}\left[1+\alpha^{|n-m|} O\left(\frac{1}{|n-m|}\right)\right] \quad \text{for } |n-m| \to \infty$$

$$\Psi_0^* S_n^y S_m^y \Psi_0 \simeq \sqrt{1-\alpha^2}\ \frac{\alpha^{|n-m|}}{|n-m|} \quad \text{for } |n-m| \to \infty$$

and for $\alpha = 1$

$$\Psi_0^* S_n^x S_m^x \Psi_0 = \Psi_0^* S_n^y S_m^y \Psi_0 \simeq \frac{1}{\sqrt{|n-m|}} \quad \text{for } |n-m| \to \infty$$

In the anisotropic case $\alpha < 1$ the Ising part $-2J^{\perp}\sum_n S_n^x S_{n+1}^x$, in (4.2) dominates and the model displays long range order in the ground state, i.e., $\Psi_0^* S_n^x S_m^x \Psi_0 \to$ const. for $|n-m| \to \infty$ For $\alpha = 1$, the isotropic limit, the long range order disappears and the static ground state correlations fall off according to power laws, $\Psi_0^* S_n^z S_m^z \Psi_0 \to |n-m|^{-2}$ and $\Psi_0^* S_n^x S_m^x \Psi_0 \simeq \Psi_0^* S_n^y S_m^y \Psi_0 \to |n-m|^{-1/2}$.

As far as the dynamical properties of the XY model are concerned, Niemeyer[42] has shown that the time-dependent longitudinal correlation function $\langle S_n^z(t) S_m^z(0)\rangle$ has the same form as the density correlations for a non-interacting Fermi gas with single particle dispersion law

$$E(k) = -J^{\perp}\sqrt{(1+\alpha)^2\cos^2(ka) + (1-\alpha)^2\sin^2(ka)}\ .$$

The transverse time-dependent correlations $\langle S_n^{\alpha}(t) S_m^{\beta}(0)\rangle$, $\alpha, \beta = x,y,z$, have been considered by McCoy et al. (ref.43).

The Heisenberg-Ising chain (4.1) has been treated by several authors (see, for instance, ref.44, where earlier references can be found). Recently, it has been shown by Sutherland[45] that the ground state properties of the model are related to the critical behaviour of the two-dimensional Baxter model[46]. Technically, the Hamiltonian (4.1) commutes with the transfer matrix for the Baxter or eight-vertex model at T_c. It follows, in particular, that the static transverse correlation function, $\Psi_0^* S_n^x S_m^x \Psi_0$ for the Heisenberg-Ising chain has the same form as the correlation function[47] for the Baxter model along a line in a square lattice. The decoupling limit, where the Baxter model reduces to two interpenetrating Ising lattices, corresponds to $J^{\parallel} = 0$, and we infer that the correlation function

$\Psi_o^{\dagger} S_n^x S_m^x \Psi_o$ for the isotropic XY model behaves as the square of the Ising correlation function at T_c. It is well-known (see, for instance, ref.48) that the critical correlations for the Ising model decay as $|n-m|^{-1/4}$ for $|n-m| \to \infty$, which is consistent with the behaviour of the transverse correlation function, $\Psi_o^{\dagger} S_n^x S_m^x \Psi_o \simeq |n-m|^{-1/2}$, for the isotropic XY model.

The dynamical properties of the Heisenberg-Ising chain at T = 0 have been considered by Luther and Peschel[49]. Using operator methods developed previously for the Luttinger model[50,51], these authors evaluated the time-dependent correlation functions. In ref.37 the present author, subsequently, calculated the correlation functions entirely within the framework of standard many body theory, employing methods developed in refs. 38 - 40 for the Tomonaga[52] and Wolff[53] models.

4.2 The Jordan-Wigner Transformation - The Luttinger Model

Using the length condition for spin half, $S^z = S^+S^- - \frac{1}{2}$, in order to eliminate S^z, and introducing $S^{\pm} = S^x \pm iS^y$, the Hamiltonian (4.1) takes a form,

$$H = -\sum_n \left[J^{\perp}(S_n^+ S_{n+1}^- + S_{n+1}^+ S_n^-) + 2J^{\parallel}(S_n^+ S_n^- - \tfrac{1}{2})(S_{n+1}^+ S_{n+1}^- - \tfrac{1}{2}) \right], \quad (4.3)$$

which clearly illustrates the many body character of the Heisenberg-Ising chain. The application of standard field theoretical methods is, however, rendered difficult because of the mixed communation relations for the spin operators, $[S_n^-, S_m^+] = 0$ for $n \neq m$ and $\{S_n^-, S_n^+\} = 1$. The whole machinery of modern many body theory[24] or field theory[25] is essentially adapted to systems satisfying Bose or Fermi statistics. This is related to the fact that Wick's theorem, which is the basic operational tool, only holds in a practical form for bosons or fermions. We mention in passing that the functional approach by Schwinger and Martin[54,55] can be applied to spin problems[56] with some success. The present author[4] has in fact made use of such an approach in the context of the anisotropic magnetic chain discussed in Chapter II.

For a spin half problem in one dimention there exists, however, an exact non-linear transformation which relates the spin operators S_n^+ and S_n^- to a spinless fermion field Ψ_n. In contrast to the standard approximative Bose representation[57] which is based on the commuting properties of the spin algebra, i.e., $[S_n^-, S_m^+] = 0$ for $n \neq m$, the Jordan-Wigner transformation[30,31] utilises the property that the spin half operators S_n^+ and S_n^- satisfy the anti-commutation relation $\{S_n^+, S_n^-\} = 1$; the Bose properties of the spin operators at different sites are accounted for by means of an appropriate phase factor. The explicit form of the transformation is

$$S_n^- = \psi_n \exp\left(-i\pi \sum_{-\infty}^{n-1} \psi_n^\dagger \psi_n\right), \quad S^z = \psi_n^\dagger \psi_n - \frac{1}{2}, \quad (4.4)$$

where the fermion field ψ_n satisfies the usual anti-commutation relations, $\{\psi_n, \psi_m^\dagger\} = \delta_{nm}$ and $\{\psi_n, \psi_m\} = 0$. Inserting (4.4), the Hamiltonian (4.3) assumes the form

$$H = \sum_n \left(J^\perp(\psi_n^\dagger \psi_{n+1} + \psi_{n+1}^\dagger \psi_n) + 2J'' n_n n_{n+1}\right), \quad (4.5)$$

where we have introduced the fermion density $n_n = \psi_n^\dagger \psi_n - \frac{1}{2}$. In a running wave basis

$$\psi_n = \int_{-\pi}^{\pi} a(k) \, e^{ikn} \frac{dk}{\sqrt{2\pi}}$$

$$n_n = \int_{-\pi}^{\pi} \rho(k) \, e^{ikn} \frac{dk}{2\pi}$$

where $\{a(k), a^\dagger(p)\} = \delta(k-p)$, $\{a(k), a(p)\} = 0$, we have

$$H = \int_{-\pi}^{\pi} (-2J^\perp \cos k) \, a^\dagger(k) \, a(k) \, dk + \int_{-\pi}^{\pi} (-2J'' \cos k) \rho(k) \rho(-k) \frac{dk}{2\pi}$$

and we conclude that the Heisenberg-Ising chain (4.1) has the same spectrum as a one-dimensional spinless Fermi gas with the single particle dispersion law $-2J^\perp \cos k$ and the two-body potential $-2J'' \cos k$.

In the absence of a magnetic field $\psi_0^* S_n^z \psi_0 = 0$. Consequently,

$$\psi_0^* \psi_n^\dagger \psi_n \psi_0 = \int_{-\pi}^{\pi} \psi_0^* a^\dagger(k) a(k) \psi_0 \frac{dk}{2\pi} = \frac{1}{2},$$

and the Fermi band shown in Fig. 4.1 is only half-filled. Owing to the Pauli principle the excitations inside the Fermi sea are dynamically "frozen". In the long wavelength-low frequency limit the dynamics of the Heisenberg-Ising chain is essentially governed by the excitations in the vicinity of the Fermi points $k_F = \pi/2$ and $k_F = -\pi/2$. It is convenient to introduce local Fock spaces[25,38] about each Fermi point and define new "smeared" or "coarse grained" fermion fields ψ_R and ψ_L for the "right" and "left" Fermi points, respectively, by

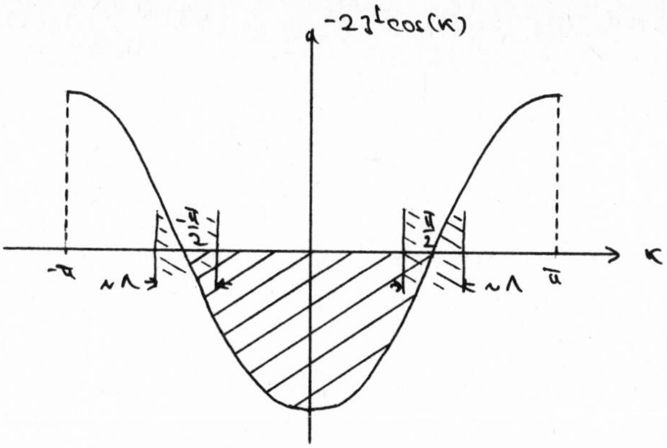

Fig.4.1 The dispersion law and the half-filled Fermi band characterising the dynamics of the isotropic XY model

$$\psi_R(x) = \int_{-\Lambda}^{\Lambda} a(k+\tfrac{\pi}{2})\, e^{ikx} \frac{dk}{\sqrt{2\pi}} ,$$

$$\psi_L(x) = \int_{-\Lambda}^{\Lambda} a(k-\tfrac{\pi}{2})\, e^{ikx} \frac{dk}{\sqrt{2\pi}} ,$$

where the momentum cut off Λ is chosen much less than one (see Fig.4.1). In terms of ψ_R and ψ_L the original microscopic Fermi field $\psi_n \simeq \psi(x)$ and density $n_n \simeq n(x)$ are given by

$$\psi(x) = i^x \psi_R(x) + i^{-x} \psi_L(x) \tag{4.7a}$$

$$n(x) = n_R(x) + n_L(x) + (-1)^x [\psi_R^\dagger(x)\psi_L(x) + \psi_L^\dagger(x)\psi_R(x)] \tag{4.7b}$$

$$n_{R(L)}(x) = \psi^\dagger_{R(L)}(x)\, \psi_{R(L)}(x) - \frac{1}{4} . \tag{4.7c}$$

The two-body interaction -2J" cosk is shown in Fig.4.2

In the weak coupling-long wavelength-low frequency limit only two kinds of scattering processes due to the two-body potential contribute (see Fig.4.2), namely the direct process $-2J''\psi_R^\dagger\psi_R\psi_L^\dagger\psi_L$ and the exchange process $2J''\psi_L^\dagger\psi_R\psi_R^\dagger\psi_L$. The processes $-2J''\psi_R^\dagger\psi_R\psi_R^\dagger\psi_R$ and $-2J''\psi_L^\dagger\psi_L\psi_L^\dagger\psi_L$ are suppressed by the Pauli principle. In Fig.4.3 we have shown the fermion interactions which dominate the dynamics of the Heisenberg-Ising chain. For a spinless fermion field we have, however, the simple crossing symmetry $\psi_L^\dagger\psi_R\psi_R^\dagger\psi_L \simeq -\psi_R^\dagger\psi_R\psi_L^\dagger\psi_L$ which enables us to include the exchange processes in the direct ones. Inserting (4.7) in (4.5), using the above crossing symmetry, and only keeping the

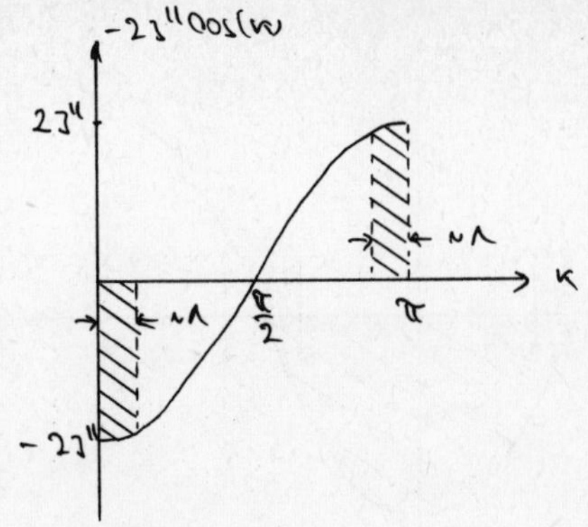

Fig.4.2 The two-body interaction -2J"cosk

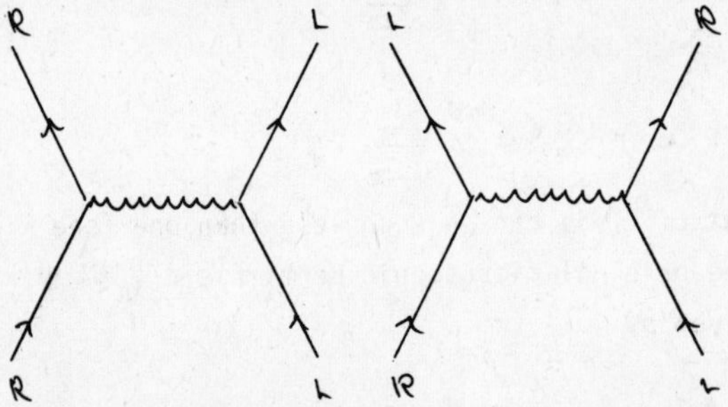

Fig.4.3 The scattering processes contributing to the dynamics of the Heisenberg-Ising chain

slowly varying terms, we obtain the effective "coarse grained" fermion Hamiltonian

$$H = \int [(-2J^{\perp})\, i(\psi_R^{\dagger}\frac{d\psi_R}{dx} - \psi_L^{\dagger}\frac{d\psi_L}{dx}) + (-8J'')n_R n_L]dx\,, \tag{4.8}$$

characterising the weak coupling-long wavelength-low frequency properties of the Heisenberg.Ising chain.

The effective Hamiltonian (4.8) is, however, just the x-space version of the Luttinger model[51] with an ultraviolet cut off prescription, which can be solved analytically by a variety of techniques[50,58-60]. The evaluation of the spin correlation functions, on the other hand, is nontrivial owing to the non linear phase factor in the Jordan-Wigner transformation (4.4). In terms of $\psi_{R(L)}$ and $n_{R(L)}$ the spin operators S^- and S^z are given by the expressions

$$S^-(x) = [i^x \psi_R(x) + i^{-x}\psi_L(x)]\, e^{-i\pi \int_{-\infty}^{x-1} (n_R(y) + n_L(y) + \frac{1}{2})dy} \tag{4.9a}$$

$$S^z(x) = n_R(x) + n_L(x) + (-1)^x [\psi_R^+(x)\psi_L(x) + \psi_L^+(x)\psi_R(x)]. \tag{4.9b}$$

In contrast to the operator approach of Luther and Peschel[49], we are not performing a continuum limit of the lattice. In order to describe the long wavelength properties of the model we have, however, assumed that the fields $\psi_R(x)$ and $\psi_L(x)$ vary slowly over a lattice distance. The effective fermion Hamiltonian (4.8) together with the Jordan-Wigner transformation (4.9) constitute a "coarse grained" description of the weak coupling-long wavelength-low frequency limit of the original microscopic Heisenberg-Ising model (4.1).

4.3 Correlation Functions for the Isotropic XY Model

In order to illustrate the field theoretical approach expounded in ref.37 in a simple case we consider first the isotropic XY model described by the effective Hamiltonian

$$H_{XY} = -2J^{\perp} i \int [\psi_R^+(x)\frac{d\psi_R(x)}{dx} - \psi_L^+(x)\frac{d\psi_L(x)}{dx}]dx. \tag{4.10}$$

In wavenumber space, defining $a_R(k) = a(k+\pi)$ and $a_L(k) = a(k-\pi)$, (4.10) takes the form

$$H_{XY} = 2J^{\perp} \int [k a_R^+(k) a_R(k) - k a_L^+(k) a_L(k)] \frac{dk}{2\pi}. \tag{4.11}$$

The single particle spectrum shown in Fig.4.4 has two branches corresponding to the propagation of "right" and "left" fermions, according to the linear dispersion laws $\omega = v(k-\pi)$ and $\omega = v(k+\pi)$, respectively. The Fermi or phase velocity $v = 2J^{\perp}$. The particle-hole or density fluctuation spectrum of $n = n_R + n_L + (-1)^x(\psi_R^+\psi_L + \psi_L^+\psi_R)$ shown in Fig.4.5 is easily inferred from Eq.(4.11) or by inspection from Fig.4.4. The spectrum is given by $\omega = \pm vk$, $\omega^2 \geq v^2(k-\pi)^2$ and $\omega^2 \geq v^2(k-\pi)^2$ for $\omega \geq 0$. Owing to the linear dispersion laws the Heisenberg equations of motion for the fields $\psi_{R(L)}$ and densities $n_{R(L)} = \psi_{R(L)}^+\psi_{R(L)} - \frac{1}{4}$ take a particular simple form,

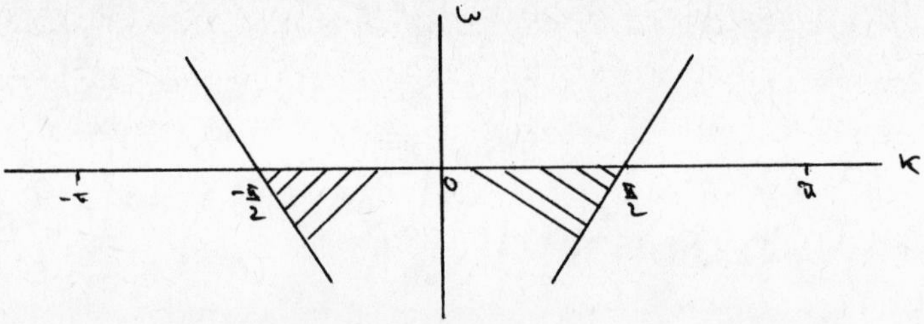

Fig.4.4 The single particle spectrum

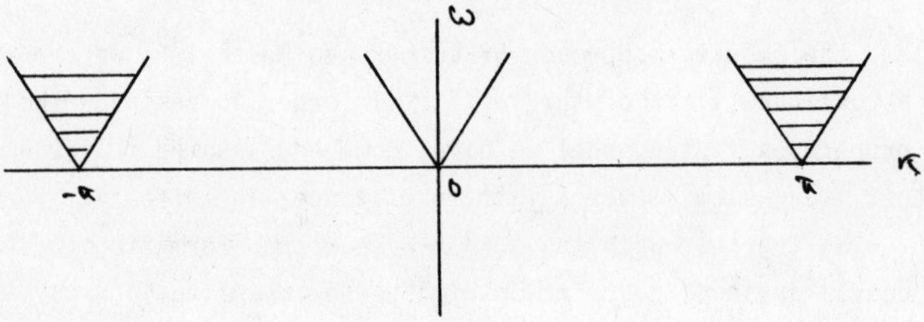

Fig.4.5 The particle-hole spectrum

$$\left(\frac{d}{dt} \overset{+}{(-)} v \frac{d}{dx}\right) \psi_{R(L)}(xt) = 0 \tag{4.12a}$$

$$\left(\frac{d}{dt} \overset{+}{(-)} v \frac{d}{dx}\right) n_{R(L)}(xt) = 0 \tag{4.12b}$$

showing that $\psi_{R(L)}$ and $n_{R(L)}$ only depend on $x \overset{(+)}{-} vt$ and thus propagate as permanent profile configurations with phase velocities $\overset{+}{(-)}v$. The spin operators, however, have a more complicated structure. For instance,

$$S^-(x) = \psi(x) \exp\left(-i\pi \int_{-\infty}^{x-1} \psi^+(y)\psi(y)\,dy\right)$$

couples essentially to the multiple particle-hole spectrum shown in Fig.4.6.

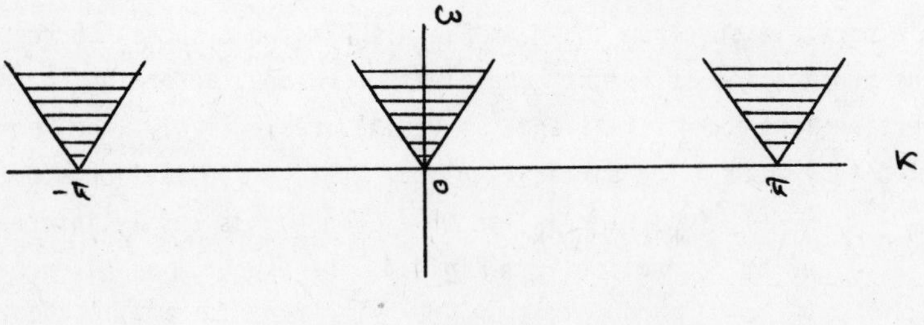

Fig.4.6 The multiple particle-hole spectrum

Let us first evaluate the static longitudinal spin correlations[31,41] characterised by $\Psi_0^* S^z(x) S^z(x')\Psi_0$. Inserting Eq.(4.9b), introducing $\Psi_0 = \Psi_{0R}\Psi_{0L}$, and using $\Psi_{0R}^* n_R \Psi_{0R} = \Psi_{0L}^* n_L \Psi_{0L} = 0$ we obtain

$$\Psi_0^* S^z(x) S^z(x') \Psi_0 =$$

$$C^0_{RR}(x,x') + C^0_{LL}(x,x') + (-1)^{x-x'}\left(P^0_{RL}(x,x') + P^0_{LR}(x',x)\right)$$

where

$$C^0_{RR}(x,x') = \Psi_{0R}^* n_R(x) n_R(x') \Psi_{0R}$$

$$C^0_{LL}(x,x') = \Psi_{0L}^* n_L(x) n_L(x) \Psi_{0L}$$

$$P^0_{RL}(x,x') = \Psi_0^* \psi_R^\dagger(x) \psi_L(x) \psi_L^\dagger(x') \psi_R(x') \Psi_0$$

$$P^0_{LR}(x',x) = \Psi_0^* \psi_L^\dagger(x') \psi_R(x') \psi_R^\dagger(x) \psi_L(x) \Psi_0 .$$

The leading behaviour for $|x - x'| \gg \Lambda^{-1}$ is determined by $C^0_{RR} + C^0_{LL}$; the next leading behaviour by $P^0_{RL} + P^0_{LR}$. By means of Wick's theorem[24,25] we have

$$C^0_{RR}(x,x') = - C^0_R(x,x')\, C^0_R(x,x')$$

$$C^0_{LL}(x,x') = - C^0_L(x,x')\, C^0_L(x,x')$$

$$P^0_{RL}(x,x') = - C^0_L(x,x')\, C^0_R(x',x)$$

$$P^0_{LR}(x',x) = - C^0_R(x',x)\, C^0_L(x,x') ,$$

where

$$C^0_R(x,x') = \Psi_{0R}^* \psi_R(x) \psi_R^\dagger(x') \Psi_{0R}$$

$$C^0_L(x,x') = \Psi_{0L}^* \psi_L(x) \psi_L^\dagger(x') \Psi_{0L}$$

are the static fermion correlation functions. For $|x - x'| \gg \Lambda^{-1}$ C^0_R and C^0_L are purely imaginary and easily evaluated. We obtain

$$C^0_R(x,x') = \int \Psi_{0R}^* a_R(k) a_R^\dagger(k) \Psi_{0R}\, e^{ik(x-x')} \frac{dk}{2\pi} =$$

$$\int_0^\Lambda e^{ik(x-x')} \frac{dk}{2\pi} = \frac{1}{2\pi} \frac{i}{x-x'} \qquad |x-x'| \gg \Lambda^{-1}$$

and, similarly, $C^0_L(xx') = -i/2\pi(x - x')$. Collecting the terms together we find in agreement with refs. 31 and 41

$$\Psi_0^* S^z(x) S^z(x') \Psi_0 = -\frac{1}{2\pi^2} \frac{1}{(x-x')^2} \left(1 - (-1)^{x-x'}\right).$$

The time dependent longitudinal correlations are easily found by noting that according to (4.12a-b) $\Psi_{R(L)}(xt) = \Psi_{R(L)}(x \mp vt)$. Consequently,

$$C^0_{RR}(xt, x't') = C^0_{RR}(x-vt, x'-vt')$$

$$C^0_{LL}(xt, x't') = C^0_{LL}(x+vt, x'+vt')$$

$$P^0_{RL}(xt, x't') = -C^0_L(x+vt, x'+vt')\, C^0_R(x'-vt', x-vt)$$

$$P^0_{LR}(x't', xt) = -C^0_R(x'-vt', x-vt)\, C^0_L(x+vt, x'+vt')$$

and we obtain

$$\Psi_0^* S^z(xt) S^z(x't') \Psi_0 =$$

$$-\frac{1}{2\pi^2} \frac{1}{(x-x')^2 - v^2(t-t')^2} \left[\frac{(x-x')^2 + v^2(t-t')^2}{(x-x')^2 - v^2(t-t')^2} - (-1)^{x-x'} \right].$$

The evaluation of the transverse spin correlation function is more complicated owing to the non-linear form of the Jordan-Wigner transformation (4.9a) which couples the spin operator S^- to the multiple particle-hole spectrum and thus introduces genuine many body effects. In the static limit[41] the calculation is in fact rather formidable involving the asymptotic properties of Toeplitz determinants. It is a peculiar feature of a quantum mechanical calculation that time is involved in an intrinsic manner. This is, of course, associated with the fact that the basic eigenvalue problem $H\Psi = E\Psi$ determines the time evolution of the states. Admitting that we have not fully assimilated the "deeper" reasons, it does turn out that in the evaluation of the transverse correlations enormous simplification is achieved by at the outset focussing on the time dependence of the correlations.

Instead of calculating the correlation function $\Psi_0^* S^-(xt) S^+(x't') \Psi_0$, characterising the transverse spin fluctuations, directly, we consider, as is customary in a many body calculation[24,55], the time-ordered Green's function

$$G^0_\perp(xt, x't') = \Psi_0^* \left(S^-(xt) S^+(x't')\right)_+ \Psi_0,$$

where the fermion fields Ψ_R and Ψ_L in the definitions of S^- and S^+ are arranged in chronologically decreasing order from left to right with a factor minus one entering each time two fermion operators are commuted, for instance, $(\psi(t)\psi^{\dagger}(t'))_{+} = \psi(t)\psi^{\dagger}(t')$ for $t > t'$ and $= -\psi^{\dagger}(t')\psi(t)$ for $t' > t$. Inserting Eq.(4.9a), noting that n commutes with ψ under time ordering, and absorbing the phase factors in the Jordan-Wigner transformations in an effective one-body potential

$$V(x''t'') = \pi\delta(t''-t)\eta(x-1-x'') - \pi\delta(t''-t')\eta(x'-1-x'')$$

we can express $G^{o}_{\perp}$ in the form

$$G^{o}_{\perp}(xt,x't') =$$

$$[G^{V}_{R}(xt,x't') + (-1)^{x-x'}G^{V}_{L}(xt,x't')]Z^{V}_{R}(xt,x't')Z^{V}_{L}(xt,x't'),$$

where

$$G^{V}_{R}(xt,x't')Z^{V}_{R}(xt,x't') =$$

$$\Psi^{*}_{OR}(\exp(-i\int dx''dt''V(x''t'')n_{R}(x''t''))\psi_{R}(xt)\psi^{\dagger}(x't'))_{+}\Psi_{OR}$$

$$G^{V}_{L}(xt,x't')Z^{V}_{L}(xt,x't') =$$

$$\Psi^{*}_{OL}(\exp(-i\int dx''dt''V(x''t'')n_{L}(x''t''))\psi_{L}(xt)\psi^{\dagger}_{L}(x't'))_{+}\Psi_{OL}$$

$$Z^{V}_{R}(xt,x't') = \Psi^{*}_{OR}(\exp(-i\int dx''dt''V(x''t'')n_{R}(x''t'')))_{+}\Psi_{OR}$$

$$Z^{V}_{L}(xt,x't') = \Psi^{*}_{OL}(\exp(-i\int dx''dt''V(x''t'')n_{L}(x''t'')))_{+}\Psi_{OL}$$

The time-ordered ground state expectation values Z^{V}_{R} and Z^{V}_{L} are easily evaluated. Since n_R and n_L obey the equations of motion (4.12b) they are essentially free Bose fields with the dispersion laws $\omega = vk$ and $\omega = -vk$. Applying Wick's theorem in generator form[24,25] to time-orderes products of Bose operators we obtain

$$Z^{V}_{R} =$$

$$\exp(-\tfrac{1}{2}\int dx_1dt_1dx_2dt_2V(x_1t_1)\Psi^{*}_{OR}(n_R(x_1t_1)n_R(x_2t_2))_{+}\Psi_{OR}V(x_2t_2))$$

$$Z^{V}_{L} =$$

$$\exp(-\tfrac{1}{2}\int dx_1dt_1dx_2dt_2V(x_1t_1)\Psi^{*}_{OL}(n_L(x_1t_1)n_L(x_2t_2))_{+}\Psi_{OL}V(x_2t_2)).$$

Furthermore, by Wick's theorem,

$$\Psi^*_{0R}(n_R(xt)\,n_R(x't'))_+\Psi_{0R} = -G^0_R(xt,x't')\,G^0_R(x't',xt)$$

$$\Psi_{0L}(n_L(xt)\,n_L(x't'))_+\Psi_{0L} = -G^0_L(xt,x't')\,G^0_L(x't',xt) \;,$$

where

$$G^0_R(xt,x't') = \Psi^*_{0R}(\psi_R(xt)\,\psi^+_R(x't'))_+\Psi_{0R}$$

$$G^0_L(xt,x't') = \Psi^*_{0L}(\psi_L(xt)\,\psi^+_L(x't'))_+\Psi_{0L}$$

are the time-ordered fermion Green's functions. Noticing that $\psi_{R(L)}(xt) = \psi_{R(L)}(x \underset{(+)}{-} vt)$ we have in the asymptotic regions off the "light cone" $(x-x')^2 - v^2(t-t')^2 = 0$ (for details see ref.37)

$$G^0_R(xt,x't') = C^0_R(x-vt,x'-vt') = \frac{i}{2\pi}\,\frac{1}{x-x'-v(t-t')}$$

$$G^0_L(xt,x't') = C^0_L(x+vt,x'+vt') = -\frac{i}{2\pi}\,\frac{1}{x-x'+v(t-t')}$$

Inserting the potential V(x"t") and integrating we obtain

$$Z^V_R(xt,x't') = \exp\left(-\frac{1}{4}\ln\left|\frac{x-x'-v(t-t')}{\xi}\right|\right) = \left|\frac{\xi}{x-x'-v(t-t')}\right|^{\frac{1}{4}}$$

$$Z^V_L(xt,x't') = \exp\left(-\frac{1}{4}\ln\left|\frac{x-x'+v(t-t')}{\xi}\right|\right) = \left|\frac{\xi}{x-x'+v(t-t')}\right|^{\frac{1}{4}} ,$$

where ξ here and in the following is a space cut off of order Λ^{-1}.

The Green's functions G^V_R and G^V_L can be derived in a heuristic manner by means of a simple gauge transformation (for a more careful derivation we refer to ref.37, see also ref.38). The gauged fermion fields

$$\overline{\psi}_{R(L)}(xt) = \psi_{R(L)}(xt)\exp(-i\,S_{R(L)}(xt))$$

satisfy the equation of motion

$$\left(\frac{d}{dt} \underset{(-)}{+} v\frac{d}{dx}\right)\overline{\psi}_{R(L)}(xt) = -i\,\overline{\psi}_{R(L)}(xt)\left(\frac{d}{dt} \underset{(-)}{+} v\frac{d}{dx}\right)S_{R(L)}(xt) ,$$

corresponding to adding a term

$$\int dx\, n_{R(L)}(xt)\left(\frac{d}{dt}\overset{+}{(-)} v\frac{d}{dx}\right)S_{R(L)}(xt),$$

i.e., a time-dependent one-body potential, to the XY Hamiltonian (4.10). Since the Green's functions G_R^W and G_L^W introduced earlier describe the propagation of "right" and "left" fermions, respectively, in an external one-body potential $W_{R(L)}$, we immediately infer the expressions

$$G_R^W(xt,x't') = G_R^0(xt,x't')\exp(-i(S_R(xt)-S_R(x't')))$$

$$G_L^W(xt,x't') = G_L^0(xt,x't')\exp(-i(S_L(xt)-S_L(x't'))),$$

where $W_{R(L)}(xt) = (\frac{d}{dt}\overset{+}{(-)}v\frac{d}{dx})S_{R(L)}(xt)$. Noticing, furthermore, that

$$\left(\frac{d}{dt}\overset{+}{(-)} v\frac{d}{dx}\right)G_{R(L)}^0(xt,x't') = \delta(x-x')\delta(t-t')$$

we have

$$S_{R(L)}(xt) = \int G_{R(L)}^0(xt,x't')\,W_{R(L)}(x't')\,dx'dt',$$

and we obtain the general solutions for arbitrary time and space dependent potentials W_R and W_L

$$G_R^W(xt,x't') =$$
$$G_R^0(xt,x't')\exp\left(-i\int dx''dt''\,W_R(x''t'')(G_R^0(xt,x''t'')-G_R^0(x't',x''t''))\right)$$
$$G_L^W(xt,x't') =$$
$$G_L^0(xt,x't')\exp\left(-i\int dx''dt''\,W_L(x''t'')(G_L^0(xt,x''t'')-G_L^0(x't',x''t''))\right).$$

Substituting the particular one-body potential V, corresponding to the Jordan-Wigner phase factors, we find $G_R^V(xt,x't') \simeq 1/\xi$ and $G_L^V(xt,x't') \simeq -\xi/(x-x'+v(t-t'))^2$. Choosing the time ordering $t > t'$ in G_L^0 the leading correction to $\Psi_0^* S^-(xt)S^+(x't')\Psi_0$ is given by

$$Z_R^V Z_L^V G_R^V \simeq \frac{1}{\xi}\left|\frac{\xi^2}{(x-x')^2 - v^2(t-t')^2}\right|^{\frac{1}{4}}.$$

The next leading term

$$(-1)^{x-x'} Z_R^V Z_L^V G_L^V \simeq$$

$$- \frac{(-1)^{x-x'}}{\xi} \left| \frac{\xi}{x-x'+v(t-t')} \right|^2 \left| \frac{\xi^2}{(x-x')^2-v^2(t-t')^2} \right|^{\frac{1}{4}}$$

is not invariant under the parity transformation x-x' → -(x-x'). This is related to the spatial asymmetry of the Jordan-Wigner transformation and is a somewhat subtle point which has been discussed in some detail in refs.43 and 49. Here, as in ref.37, we circumvent the difficulty in a somewhat ad hoc manner by simply symmetrising the next leading term. We thus obtain

$$\Psi_0^* S^-(xt) S^+(x't')\Psi_0 \simeq$$

$$\frac{1}{\xi} \left| \frac{\xi^2}{(x-x')^2-v^2(t-t')^2} \right|^{\frac{1}{4}} \left[1-(-1)^{x-x'} \frac{\xi^2}{(x-x')^2-v^2(t-t')^2} \right].$$

Summarising and inserting missing details (see ref.37) we have for the isotropic XY model the longitudinal and transverse ground state correlations

$$\Psi_0^* S^z(xt) S^z(00)\Psi_0 = -\frac{1}{2\pi^2} \frac{1}{x^2-v^2t^2} \left[\frac{x^2+v^2t^2}{x^2-v^2t^2} - (-1)^x \right]$$

$$\text{for } |x^2-v^2t^2| \gg \Lambda^{-2}, \quad v = 2J^\perp \tag{4.13}$$

and

$$\Psi_0^* S^-(xt) S^+(00)\Psi_0 = \frac{1}{\xi} \left[\left| \frac{\xi^2}{x^2-v^2t^2} \right|^{\frac{1}{4}} \exp(i\frac{\pi}{4}\,\mathrm{sign}(t)\,\eta(v^2t^2-x^2)) \right.$$

$$\left. - (-1)^x \left| \frac{\xi^2}{x^2-v^2t^2} \right|^{\frac{5}{4}} \exp(i\frac{5\pi}{4}\,\mathrm{sign}(t)\,\eta(v^2t^2-x^2)) \right]$$

$$\text{for } |x^2-v^2t^2| \gg \Lambda^{-2}, \quad v = 2J^\perp \tag{4.14}$$

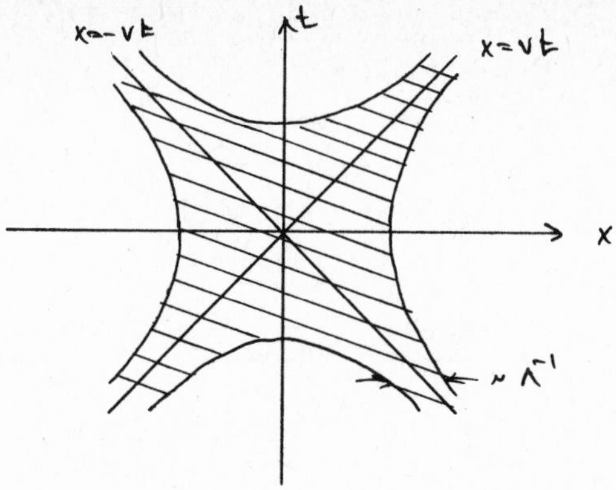

Fig.4.7 The "light cone" for the isotropic XY model. The asymptotic expressions are valid in the unshaded regions

in conformity with refs.42 and 49. The asymptotic approximations used in deriving the expressions (4.13) and (4.14) are valid in the regions $|x^2 - v^2t^2| \gg \Lambda^{-2}$ off the "light cone" $x^2 - v^2t^2 = 0$, as indicated in Fig.4.7 (for details we refer to ref.37).

In the static limit t = 0 we have

$$\Psi_0^* S^z(x) S^z(0) \Psi_0 = -\frac{1}{2\pi^2}\frac{1}{x^2}\left(1-(-1)^x\right) \tag{4.15}$$

$$\Psi_0^* S^-(x) S^+(0) \Psi_0 = \frac{1}{\xi}\left|\frac{\xi}{x}\right|^{\frac{1}{2}}\left[1-(-1)^x\left|\frac{\xi}{x}\right|^2\right] \tag{4.16}$$

for $|x| \gg \Lambda^{-1}$,

in accordance with refs. 31 and 41. The longitudinal and transverse ground state correlations for the isotropic XY model thus fall off as $1/x^2$ and $1/\sqrt{x}$, respectively, i.e., there is no long range order. As noted in ref.31 the expression (4.15) for the longitudinal correlations actually holds for all values of x on the one-dimensional lattice, x = n, n = 1,2,... The transverse correlation function (4.16), on the other hand, is only valid asymptotically for $|x| \gg \Lambda^{-1}$. The space cut off ξ is of order Λ^{-1} and is only qualitatively known. The approach by McCoy[47] using the asymptotic properties of Toeplitz determinants enables one to determine ξ. McCoy finds

$$\Psi_0^* S^-(x) S^+(0) \Psi_0 = 2e^{\frac{1}{2}} 2^{\frac{2}{3}} A^{-6} \frac{1}{\sqrt{x}}\left[1-(-1)^x\frac{1}{8x^2}\right],$$

where A = 1.2824.. is Glaisher's constant.

4.4 The Spectrum of the Isotropic XY Model

The physical properties of the isotropic XY model are most easily discussed in terms of the transverse and longitudinal magnetic susceptibilities $\chi_\perp(k,\omega)$ and $\chi_\parallel(k,\omega)$. We have, as in Chapter II,

$$\chi^0_{\perp,\parallel}(k,z) = \int_{-\infty}^{\infty} \frac{\chi''^0_{\perp,\parallel}(k,\omega)}{\omega - z}\,\frac{d\omega}{\pi} ,$$

where z is complex, $\chi^0_{\perp,\parallel}(k,\omega) = \lim_{\varepsilon\to 0} \chi^0_{\perp,\parallel}(k,\omega + i\varepsilon)$. The spectral functions $\chi^0_\perp(k,\omega)$ and $\chi''^0_\parallel(k,\omega)$ are given by

$$\chi''^0_\parallel(k,\omega) = \int_{-\infty}^{\infty} \tfrac{1}{2}\Psi_0^*[S^z(x,t),S^z(x',t')]\Psi_0\, e^{i\omega(t-t')-ik(x-x')}\,d(x-x')\,d(t-t')$$

$$\chi''^0_\perp(k,\omega) = \int_{-\infty}^{\infty} \tfrac{1}{2}\Psi_0^*[S^-(x,t),S^+(x',t')]\Psi_0\, e^{i\omega(t-t')-ik(x-x')}\,d(x-x')\,d(t-t').$$

From the expressions (4.13) and (4.14) for the longitudinal and transverse correlations we obtain the corresponding spectral functions (for details, see ref.37)

$$\chi''^0_\parallel(k,\omega) = vk^2\delta(\omega^2 - v^2k^2) + \eta(\omega^2 - v^2(k-\pi)^2) + \eta(\omega^2 - v^2(k+\pi)^2) \qquad (4.17)$$

$$\chi''^0_\perp(k,\omega) = \left[\frac{\Omega^2}{\omega^2 - v^2k^2}\right]^{\frac{3}{4}} \eta(\omega^2 - v^2k^2) + \left[\frac{\Omega^2}{\omega^2 - v^2(k-\pi)^2}\right]^{-\frac{1}{4}} \eta(\omega^2 - v^2(k-\pi)^2) + \left[\frac{\Omega^2}{\omega^2 - v^2(k+\pi)^2}\right]^{-\frac{1}{4}} \eta(\omega^2 - v^2(k+\pi)^2) \qquad (4.18)$$

Here Ω is a wavenumber cut off of order Λ and the above expressions are valid for $|k| \ll \Lambda$ and $|\omega| \ll v\Lambda$

The longitudinal response has a sharp resonance at $\omega = \pm vk$, corresponding to the propagation of a longitudinal spin wave with linear dispersion and phase velocity $v = 2J^{\perp}$. This mode will be discussed in more detail in Section 4.6. The band contributions at $k = \pi$ and $k = -\pi$ are associated with the short range order. As the lattice shrinks to zero they move out to infinity. The longitudinal response can be interpreted in terms of the particle-hole spectrum of the underlying Fermi model (4.11). The spin waves at $\omega = \pm vk$ correspond to the long wavelength Bose-like density fluctuations, the so-called Tomonaga Bosons[38,50,52], whereas the bands arise from the particle-hole continuum contributions at $k = \pm\pi$ (see Fig.4.5). In Fig.4.8 we have shown the longitudinal response $\chi^{0\,\prime\prime}_{\parallel}(k,\omega)$ as a function of the wavenumber k.

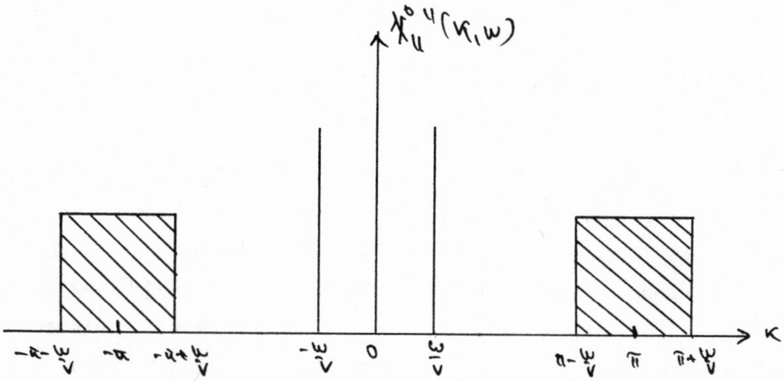

Fig.4.8 The longitudinal response of the isotropic XY model as a function of the wavenumber k for fixed frequency ω (arbitrary units)

The transverse response has band contributions both in the long wavelength limit k = 0 and at the zone edges $k = \pm\pi$; consequently, there are no transverse spin waves. In terms of the underlying Fermi model the bands are due to the coupling of the transverse spin operator

$$S^{-}(x) = \psi(x)\exp\left(-i\pi\int_{-\infty}^{x-1}\psi^{+}(y)\psi(y)dy\right)$$

to the multiple particle-hole continuum (see Fig.4.6). We also remark that the response is algebraically divergent at the band edges $k = \pm\omega/v$ with the exponent -3/2; at the band edges $k = \pi\pm\omega/v$ and $k = -\pi\pm\omega/v$ the response vanishes with the power 1/2. In Fig.4.9 we have shown the transverse response $\chi^{0\,\prime\prime}_{\perp}(k,\omega)$ as a function of the wavenumber k.

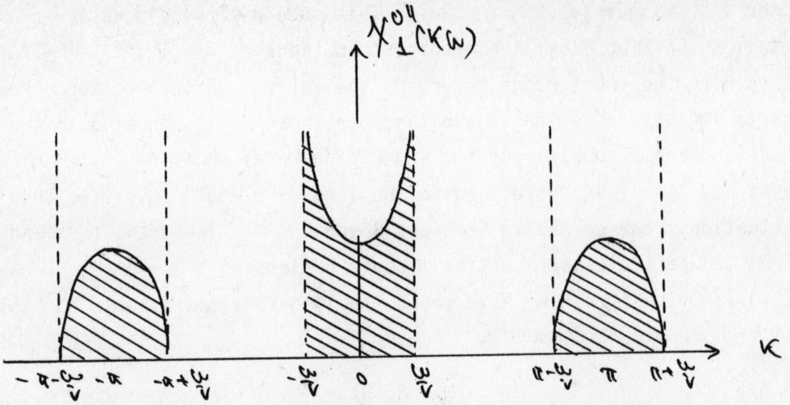

Fig.4.9 The transverse response of the isotropic XY model as a function of k for fixed ω (arbitrary units)

4.5 Correlation Functions for the Heisenberg-Ising Chain

We now turn our attention to the Heisenberg-Ising chain described by the effective fermion Hamiltonian (4.8). Here the longitudinal exchange $-8J''\int dx n_R n_L$ couples the fermion excitations at opposite Fermi points and gives rise to corrections to the exponents for the XY model, discussed in a previous section. Since the Luttinger model (4.8) applies to the microscopic Heisenberg-Ising chain in the weak coupling limit we can only determine the leading corrections to the exponents.

The longitudinal correlations are characterised by $\Psi_0^* S^z(xt) S^z(x't')\Psi_0$. In order to compute the corrections due to J" it is, however, convenient to introduce the time-ordered Green's function[24,55]

$$G_{\parallel}(xt,x't') = \Psi^*(S^z(xt)\, S^z(x't'))_+ \Psi ,$$

where Ψ is the ground state of (4.8),

$$H = H_{XY} - 8J''\int n_R(x)\, n_L(x)\, dx ,$$

and the longitudinal spin operator (4.9b),

$$S^z = n_R + n_L + (-1)^x(\psi_R^\dagger \psi_L + \psi_L^\dagger \psi_R),$$

develops in time according to the Heisenberg equation of motion[55] $i dS^z/dt = [S^z, H]$. In the interaction representation[24]

$$G_{11}(xt,x't') = G_{RR}(xt,x't') + G_{LL}(xt,x't') + G_{RL}(xt,x't') + G_{LR}(xt,x't') + (-1)^{x-x'}\left(P_{RL}(xt,x't') + P_{LR}(x't',xt)\right),$$

where

$$G_{RR}(xt,x't') = \frac{\Psi_0^*\left(\exp(i\delta J''\int dx\,dt\, n_R n_L)\, n_R(xt)\, n_R(x't')\right)_+ \Psi_0}{\Psi_0^*\left(\exp(i\delta J''\int dx\,dt\, n_R n_L)\right)_+ \Psi_0}$$

$$G_{LL}(xt,x't') = \frac{\Psi_0^*\left(\exp(i\delta J''\int dx\,dt\, n_R n_L)\, n_L(xt)\, n_L(x't')\right)_+ \Psi_0}{\Psi_0^*\left(\exp(i\delta J''\int dx\,dt\, n_R n_L)\right)_+ \Psi_0}$$

$$G_{RL}(xt,x't') = \frac{\Psi_0^*\left(\exp(i\delta J''\int dx\,dt\, n_R n_L)\, n_R(xt)\, n_L(x't')\right)_+ \Psi_0}{\Psi_0^*\left(\exp(i\delta J''\int dx\,dt\, n_R n_L)\right)_+ \Psi_0}$$

$$P_{RL}(xt,x't') = \frac{\Psi_0^*\left(\exp(i\delta J''\int dx\,dt\, n_R n_L)\, \psi_R^\dagger(xt)\, \psi_L(xt)\, \psi_L^\dagger(x't')\, \psi_R(x't')\right)_+ \Psi_0}{\Psi_0^*\left(\exp(i\delta J''\int dx\,dt\, n_R n_L)\right)_+ \Psi_0}$$

$$G_{RL}(xt,x't') = G_{LR}(x't',xt) \quad \text{and} \quad P_{RL}(xt,x't') = P_{LR}(x't',xt)$$

To first order in J" only G_{RL} and G_{LR} contribute to the leading behaviour of G_{11}. Expanding the exponential and applying Wick's theorem[24,25] we obtain

$$G_{RL}(xt,x't') = i\delta J''\int dx''dt''\, G^0_{RR}(xt,x''t'')\, G^0_{LL}(x''t'',x't'),$$

where

$$G^0_{RR}(xt,x't') = \Psi_0^*\left(n_R(xt)\, n_R(x't')\right)_+ \Psi_0$$

$$G^0_{LL}(xt,x't') = \Psi_0^*\left(n_L(xt)\, n_L(x't')\right)_+ \Psi_0$$

are the time ordered density correlation functions for the XY model. Again by Wick's theorem,

$$G^0_{RR}(xt,x't') = G^0_R(xt,x't')^2 \quad \text{and} \quad G^0_{LL}(xt,x't') = G^0_L(xt,x't')^2,$$

where from Section 4.3 the time ordered fermion correlation functions

$$G^0_R(xt,x't') = \Psi_0^*(\psi_R(xt)\psi_R^\dagger(x't'))_+\Psi_0$$

$$G^0_L(xt,x't') = \Psi_0^*(\psi_L(xt)\psi_L^\dagger(x't'))_+\Psi_0$$

are given by

$$G^0_R(xt,x't') = \frac{i}{2\pi(x-x'-v(t-t'))}$$

$$G^0_L(xt,x't') = \frac{-i}{2\pi(x-x'+v(t-t'))} .$$

In frequency-wavenumber space

$$G^0_R(k,\omega) = \frac{i}{\omega - vk}, \quad G^0_L(k,\omega) = \frac{i}{\omega + vk}$$

and, consequently, by differentiation

$$G^0_{RR}(k,\omega) = \frac{ik}{2\pi(\omega - vk)}, \quad G^0_{LL}(k,\omega) = \frac{-ik}{2\pi(\omega + vk)} .$$

Furthermore, the leading correction takes the form

$$2G_{RL}(k,\omega) = 16iJ'' G^0_{RR}(k,\omega)\, G^0_{LL}(k,\omega) .$$

Rearranging and inserting $v = 2J^{\perp}$ we obtain, choosing the time order $t > t'$, to first order in J''

$$\Psi^* S^z(xt) S^z(x't')\Psi = -\left[1 + \frac{2J''}{\pi J^{\perp}}\right]\frac{1}{4\pi^2}\left[\left(\frac{1}{x-x'-v(t-t')}\right)^2 + \left(\frac{1}{x-x'+v(t-t')}\right)^2\right].$$

The next leading correction to $G_{\parallel}$ is given by $2(-1)^{x-x'}P_{RL}(xtx't')$. Noticing that n_R and n_L commute under time ordering and using the Stratonovich-Hubbard transformation[38,61,62]

$$\left(\exp\left(i8J''\int dx\,dt\, n_R n_L\right)\right)_+ =$$

$$\frac{\int \prod dW_R\, dW_L \exp\left(-\frac{i}{8J''}\int dx\,dt\, W_R W_L\right)\left(\exp\left(-i\int dx\,dt\,(W_R n_R + W_L n_L)\right)\right)_+}{\int \prod dW_R\, dW_L \exp\left(-\frac{i}{8J''}\int dx\,dt\, W_R W_L\right)}$$

we can express $2(-1)^{x-x'}P_{RL}(xtx't')$ in the form

$$-2(-1)^{x-x'}\frac{\int \Pi dW_R dW_L \exp\left(-\frac{i}{8J''}\int dx dt\, W_R W_L\right) Z_R^W Z_L^W G_L^W(xt,x't') G_R^W(x't',xt)}{\int \Pi dW_R dW_L \exp\left(-\frac{i}{8J''}\int dx dt\, W_R W_L\right) Z_R^W Z_L^W}.$$

The functional integral $\int \tilde{\Pi} dW$ is to be interpreted as the continuum limit of a multiple integral $\int \Pi dW(xt)$ over discretised space-time. The amplitudes Z_R^W and Z_L^W characterise the polarisation of the ground state due to the potential W . To leading order in J" we neglect these corrections and set $Z_R^W = Z_L^W = 1$. In the previous section we found

$$G_R^W(xt,x't')/G_R^0(xt,x't') =$$

$$\exp\left(-i\int dx''dt''\, W_R(x''t'')(G_R^0(xt,x''t'') - G_R^0(x't',x''t''))\right)$$

$$G_L^W(xt,x't')/G_L^0(xt,x't') =$$

$$\exp\left(-i\int dx''dt''\, W_L(x''t'')(G_L^0(xt,x''t'') - G_L^0(x't',x''t''))\right).$$

Inserting Z_R^W, Z_L^W, G_R^W, and G_L^W, and again using the Stratonovich-Hubbard transformation we obtain

$$-2(-1)^{x-x'} G_R^0(x't',xt)\, G_L^0(xt,x't') \times$$

$$\exp\left(i8J''\int dx''dt''\, (G_R^0(x't',x''t'') - G_R^0(xt,x''t''))(G_L^0(xt,x''t'') - G_L^0(x't',x''t''))\right)$$

The integral is most easily performed in Fourier space. Using $G_R^0(k,\omega) = i/(\omega - vk)$, $G_L^0(k,\omega) = i/(\omega + vk)$, and

$$\int \frac{e^{-i\omega t + ikx}}{\omega^2 - v^2k^2}\frac{dk}{2\pi}\frac{d\omega}{2\pi} = \frac{i}{4\pi v}\ln|x^2 - v^2t^2|$$

we find the next leading correction to $G_{||}$

$$(-1)^{x-x'}\left|\frac{\xi^2}{(x-x')^2 - v^2(t-t')^2}\right|^{1+\frac{2J''}{\pi J^{\perp}}}.$$

Inserting missing phase factors arising from integrating across the "light cone" (see ref.37), we finally have for the longitudinal ground state correlations of the Heisenberg-Ising chain

$$\Psi^* S^z(xt) S^z(00) \Psi = -\frac{1}{2\pi^2}\left(1+\frac{2J''}{\pi J^\perp}\right)\frac{x^2+v^2t^2}{(x^2-v^2t^2)^2}$$

$$+ (-1)^x \left|\frac{\xi^2}{x^2-v^2t^2}\right|^{1+\frac{2J''}{\pi J^\perp}} \exp(i\pi(1+\frac{2J''}{\pi J^\perp})\,\mathrm{sign}(t)\,\eta(v^2t^2-x^2))$$

$$\text{for } |x^2-v^2t^2| \gg \Lambda^{-2} \qquad (4.19)$$

The static longitudinal correlations are given by

$$\Psi^* S^z(x) S^z(0) \Psi = -\frac{1}{2\pi^2}\left(1+\frac{2J''}{\pi J^\perp}\right)\frac{1}{x^2} + (-1)^x \left|\frac{\xi}{x}\right|^{2+\frac{4J''}{\pi J^\perp}} \qquad (4.20)$$

$$\text{for } |x| \gg \Lambda^{-1}$$

The transverse correlation function $\Psi^* S^-(xt)S^+(x't')\Psi$ is more complicated to evaluate owing to the interference of the Jordan-Wigner phase factor with the two-body potential. Defining the Green's function

$$G_\perp(xt,x't') = \Psi^*(S^-(xt)\,S^+(x't'))_+\Psi,$$

introducing

$$S^-(x) = (i^x\psi_R(x)+i^{-x}\psi_L(x))\exp(-i\pi\int_{-\infty}^{x-1} dq\,(n_R+n_L+\tfrac{1}{2})),\quad S^+=(S^-)^+,$$

and transforming to the interaction representation[24], we have

$$G_\perp(xt,x't') = G_R(xt,x't') + (-1)^{x-x'}G_L(xt,x't'),$$

where

$$G_{R(L)}(xt,x't') =$$

$$\frac{\Psi_0^*(\exp(i8J''\int dx dt\, n_R n_L)\exp(-i\int dx dt\,(n_R+n_L)V)\psi_{R(L)}(xt)\psi^+_{R(L)}(x't'))_+\Psi_0}{\Psi_0^*(\exp(i8J''\int dx dt\, n_R n_L))_+\Psi_0}.$$

By means of the Stratonovich-Hubbard transformation used above we can again decouple the two channels pertaining to the right and left Fermi points, respectively,

$$G_{R(L)}(xt,x't') = \frac{\int\prod d\psi_R d\psi_L \exp(-\frac{i}{8J''}\int dx dt\,\psi_R\psi_L)Z_R^{\psi+V}Z_L^{\psi+V}G^{\psi+V}_{R(L)}(xt,x't')}{\int\prod d\psi_R d\psi_L \exp(-\frac{i}{8J''}\int dx dt\,\psi_R\psi_L)Z_R^{\psi}Z_L^{\psi}},$$

where from before

$$Z^{W}_{R(L)} = \exp\left(-\frac{1}{2}\int dx\,dt\,dx'dt'\; W(xt)\, G^{0}_{R(L)R(L)}(xt,x't')\, W(x't')\right)$$

and

$$G^{W}_{R(L)}(xt,x't')/G^{0}_{R(L)}(xt,x't') =$$

$$\exp\left(-i\int dx''dt''\; W(x''t'')\left(G^{0}_{R(L)}(xt,x''t'') - G^{0}_{R(L)}(x't',x''t'')\right)\right).$$

Expanding $Z^{W+V}_{R(L)}$ and $G^{W+V}_{R(L)}$, neglecting ground state polarisations, i.e., setting $Z^{W}_{R} = Z^{W}_{L} = 1$, and using the Stratonovich-Hubbard transformation we obtain

$$G_{R(L)}(xt,x't') = Z^{V}_{R}Z^{V}_{L}G^{V}_{R(L)}(xt,x't')\exp\Big[8J''\int dx_1dt_1dx_2dt_2\Big[\,G^{0}_{R(L)}(xt,x_1t_1)$$

$$- G^{0}_{R(L)}(x't',x_1t_1) - i\int dx_3dt_3\, V(x_3t_3)\,G^{0}_{R(L)R(L)}(x_3t_3,x_1t_1)\Big]\times$$

$$G^{0}_{R(L)}(x_1t_1,x_2t_2)\,V(x_2t_2)\Big]$$

The integrals in the exponents are easily reduced in Fourier space. Inserting $G^{0}_{R(L)}(k\omega) = i/(\omega \underset{(+)}{-} vk)$, $G^{0}_{R(L)R(L)}(k\omega) = \underset{(-)}{+} ik/2\pi(\omega \underset{(+)}{-} vk)$, and using $V(x''t'') = \bar{V}(xtx''t'') - \bar{V}(x't'x''t'')$, where $\bar{V}(xtx't') = \pi\delta(t-t')\eta(x-x')$, i.e., $\bar{V}(k\omega) = -i\pi/k$, we have

$$G_{R(L)}(xt,x't') = Z^{V}_{R}Z^{V}_{L}G^{V}_{R(L)}(xt,x't')\times$$

$$\exp\left[\underset{(-)}{+}4iJ''(2\underset{(+)}{-}1)\int\frac{d\omega}{2\pi}\frac{dk}{2\pi}\,\frac{e^{-i\omega(t-t')+ik(x-x')}}{\omega^2 - v^2k^2}\right].$$

Since

$$\int\frac{d\omega}{2\pi}\frac{dk}{2\pi}\,\frac{e^{-i\omega t+ikx}}{\omega^2 - v^2k^2} = \frac{i}{4\pi v}\ln|x^2 - v^2t^2|,$$

we obtain

$$G_R(xt,x't') = G_R^{XY}(xt,x't')\,\left|(x-x')^2 - v^2(t-t')^2\right|^{-\frac{J''}{2\pi J^{\perp}}}$$

$$G_L(xt,x't') = G_L^{XY}(xt,x't')\,\left|(x-x')^2 - v^2(t-t')^2\right|^{\frac{3J''}{2\pi J^{\perp}}},$$

where $G_R^{XY} = Z^V Z^V G^V$ and $G_L^{XY} = Z_R^V Z_L^V G_L^V$ are the corresponding results for the XY model. Inserting factors due to the "light cone" integrations (see ref.37), we finally arrive at the following expression for the transverse spin correlations of the Heisenberg-Ising chain:

$$\Psi^* S^-(xt)\, S^+(00)\Psi = \frac{1}{\xi}\Bigg[\left|\frac{\xi^2}{x^2-v^2t^2}\right|^{\frac{1}{4}-\frac{J''}{2\pi J^{\perp}}} \exp\!\left(i\pi\left(\tfrac{1}{4}-\tfrac{J''}{2\pi J^{\perp}}\right)\right)\operatorname{sign}(t)\,\eta(v^2t^2-x^2)$$

$$- (-1)^x \left|\frac{\xi^2}{x^2-v^2t^2}\right|^{\frac{3}{4}+\frac{3J''}{2\pi J^{\perp}}} \exp\!\left(i\pi\left(\tfrac{5}{4}+\tfrac{3J''}{2\pi J^{\perp}}\right)\right)\operatorname{sign}(t)\,\eta(v^2t^2-x^2)\Bigg] \tag{4.21}$$

for $|x^2-v^2t^2| \gg \Lambda^{-2}$

In the static limit we have

$$\Psi^* S^-(x)\, S^+(0)\Psi = \frac{1}{\xi}\left[\left|\frac{\xi}{x}\right|^{\frac{1}{2}-\frac{J''}{\pi J^{\perp}}} - (-1)^x\left|\frac{\xi}{x}\right|^{\frac{5}{2}+\frac{3J''}{\pi J^{\perp}}}\right] \tag{4.22}$$

for $|x| \gg \Lambda^{-1}$

A comment about the Fermi velocity v. To first order in J" the present asymptotic approach dose not yield corrections to the phase velocity $v = 2J^{\perp}$. From the microscopic fermion model (4.6) we obtain, however, the Hartree-Fock correction $-4J''/\pi$, in agreement with ref.47. In the following we assume that v is given by $2J^{\perp} - 4J''/\pi$.

4.6 The Spectrum of the Heisenberg-Ising Chain

The spectrum of the Heisenberg-Ising model is most conveniently discussed in terms of the longitudinal and transverse frequency and wave-number dependent complex susceptibilities $\chi_{\parallel}(k\omega)$ and $\chi_{\perp}(k\omega)$. Summarising, we have the spectral representation

$$\chi_{\parallel,\perp}(k,z) = \int_{-\infty}^{\infty} \frac{\chi''_{\parallel,\perp}(k,\omega)}{\omega - z}\,\frac{d\omega}{\pi},$$

where the spectral functions or dissipative responses $\chi''_{\perp}(k,\omega)$ and $\chi''_{\parallel}(k,\omega)$ are given by

$$\chi''_{\parallel}(k,\omega) = \int_{-\infty}^{\infty} \tfrac{1}{2}\Psi^{*}[S^{z}(xt), S^{z}(00)]\Psi\, e^{i\omega t - ikx}\, dt\, dx$$

$$\chi''_{\perp}(k,\omega) = \int_{-\infty}^{\infty} \tfrac{1}{2}\Psi^{*}[S^{-}(xt), S^{+}(00)]\Psi\, e^{i\omega t - ikx}\, dt\, dx$$

From the expressions (4.19) and (4.20) for the longitudinal and transverse spin correlations we obtain, using the algorithms

$$\chi_{\parallel}(k,\omega + i\varepsilon\, \mathrm{sign}(\omega)) = i\, G_{\parallel}(k,\omega)$$

$$\chi_{\perp}(k,\omega + i\varepsilon\, \mathrm{sign}(\omega)) = i\,(G_{\perp}(xt)\,\mathrm{sign}(t))(k,\omega)$$

the longitudinal and transverse spectral functions

$$\begin{aligned}\chi''_{\parallel}(k,\omega) = {} & 2J^{\perp}\left(1 - \frac{2J^{\parallel}}{\pi J^{\perp}}\right) k^{2}\,\delta(\omega^{2} - v^{2}k^{2}) \\ & + \left(\frac{\Omega^{2}}{\omega^{2} - v^{2}(k-\pi)^{2}}\right)^{-\frac{2J^{\parallel}}{\pi J^{\perp}}} \eta(\omega^{2} - v^{2}(k-\pi)^{2}) \\ & + \left(\frac{\Omega^{2}}{\omega^{2} - v^{2}(k+\pi)^{2}}\right)^{-\frac{2J^{\parallel}}{\pi J^{\perp}}} \eta(\omega^{2} - v^{2}(k+\pi)^{2})\end{aligned} \tag{4.23}$$

and

$$\begin{aligned}\chi''_{\perp}(k,\omega) = {} & \left(\frac{\Omega^{2}}{\omega^{2} - v^{2}k^{2}}\right)^{\frac{3}{4} + \frac{J^{\parallel}}{2\pi J^{\perp}}} \eta(\omega^{2} - v^{2}k^{2}) \\ & + \left(\frac{\Omega^{2}}{\omega^{2} - v^{2}(k-\pi)^{2}}\right)^{-\frac{1}{4} - \frac{3J^{\parallel}}{2\pi J^{\perp}}} \eta(\omega^{2} - v^{2}(k-\pi)^{2}) \\ & + \left(\frac{\Omega^{2}}{\omega^{2} - v^{2}(k+\pi)^{2}}\right)^{-\frac{1}{4} - \frac{3J^{\parallel}}{2\pi J^{\perp}}} \eta(\omega^{2} - v^{2}(k+\pi)^{2})\end{aligned} \tag{4.24}$$

These expressions are valid in the weak coupling-long wavelength-low frequency limit, i.e. for $|J''/J^{\perp}| \ll 1$, $k \ll \Lambda$, and $\omega \ll v\Lambda$. The phase or Fermi velocity $v = 2J^{\perp} - 4J''/\pi$ and Ω is a wavenumber cutoff of order Λ.

The longitudinal response exhibits sharp resonances at $\omega = \pm vk$, corresponding to the propagation of longitudinal spin waves with linear dispersion and phase velocities $\pm v$. We notice, however, that there is no long range order in the ground state as seen from the longitudinal correlation function (4.20), which decays at long distances. The spin wave is therefore not a Goldstone mode associated with a broken symmetry[21] but rather a consequence of the conservation law $\frac{dS^z}{dt} + \frac{dJ^z}{dx} = 0$ for the longitudinal spin fluctuations, arising from the rotational invariance of the Heisenberg-Ising chain about the z-axis. In a heuristic manner we obtain, expanding the spin current in terms of dS^z/dx, $J^z(xt) = -v^2\int^t dt' dS^z(xt')/dx$, and noticing that the ground state dynamics is time reversal invariant, a wave equation for S^z implying spin waves. In the ferromagnetic case $J''/J^{\perp} > 0$ the phase velocity $v = 2J^{\perp} - 4J''/\pi J^{\perp}$ decreases as J" grows, indicating that the spin waves become stiffer due to the enhancement of the spin alignment, and propagate more slowly. In the antiferromagnetic case $J''/J^{\perp} < 0$ the phase velocity grows with increasing $|J''/J^{\perp}|$, i.e., the spin waves become softer and propagate more rapidly. From the work of Johnson et al.[47] we can extract the exact expression for the phase velocity v ,

$$v = \pi \frac{\sqrt{(J^{\perp})^2 - (J'')^2}}{\operatorname{Arccos}\left(-\frac{J''}{J^{\perp}}\right)} \tag{4.25}$$

As we approach the isotropic ferromagnet $J'' = J^{\perp}$ the phase velocity v vanishes. The system develops long range order and the spin wave becomes a Goldstone mode with a quadratic dispersion law $\omega \propto k^2$, i.e., the next leading term in an expansion of ω in powers of k comes into play. In the case of the isotorpic antiferromagnet $J'' = -J^{\perp}$ and $v = \pi J^{\perp}$ in agreement with the work of Des Cloiseux and Pearson[63]. In Fig.4.10 we have plotted the phase velocity (4.25) and also shown the Hartree-Fock approximation $v = 2J^{\perp} - 4J''/\pi$. We notice that the linear approximation to the phase velocity is quite good except in the neighbourhood of the ferromagnetic point $J'' = J^{\perp}$.

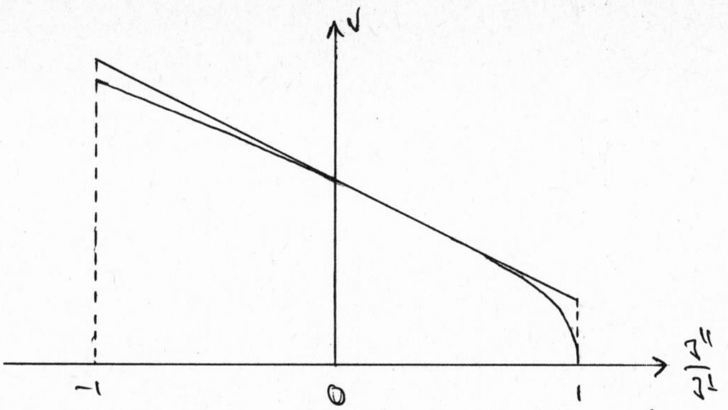

Fig.4.10 The phase velocity $v = \pi((J^{\perp})^2 - (J'')^2)^{\frac{1}{2}}/\mathrm{Arccos}(-J''/J^{\perp})$ and the Hartree-Fock approximation $v = 2J^{\perp} - 4J''/\pi$ as a function of $J''/J^{\perp}$ (arbitrary units)

The longitudinal response has continuum contributions at the zone boundaries $k = \pm\pi$ for $\omega^2 - v^2(k \pm \pi) > 0$. These bands are associated with the short range order in the ground state. In the continuum limit where the lattice shrinks to zero the bands move out to infinity. In the vicinity of the band edges the response diverges as $[\omega^2 - v^2(k\pm\pi)]^{2J''/\pi J^{\perp}}$ in the antiferromagnetic case $J''/J^{\perp} < 0$. In the ferromagnetic case $J''/J^{\perp} > 0$ the response is finite. In Figs. 4.11 and 4.12 we have shown the ferromagnetic and antiferromagnetic longitudinal response as a function of the wavenumber k for fixed frequency ω.

The static longitudinal susceptibility is given by

$$\chi_{\parallel}(\kappa) = \int_{-\infty}^{\infty} \frac{\chi''_{\parallel}(\kappa,\omega)}{\omega} \frac{d\omega}{\pi} .$$

From (4.23) we obtain

$$\chi_{\parallel}(\kappa) = \frac{1}{4\pi} \frac{1}{J^{\perp}} \left(1 + \frac{2J''}{\pi J^{\perp}}\right) + \mathrm{cst.}|k-\pi|^{\frac{4J''}{\pi J^{\perp}}} + \mathrm{cst.}|k+\pi|^{\frac{4J''}{\pi J^{\perp}}}$$

for $|k| \ll \Lambda$ (4.26)

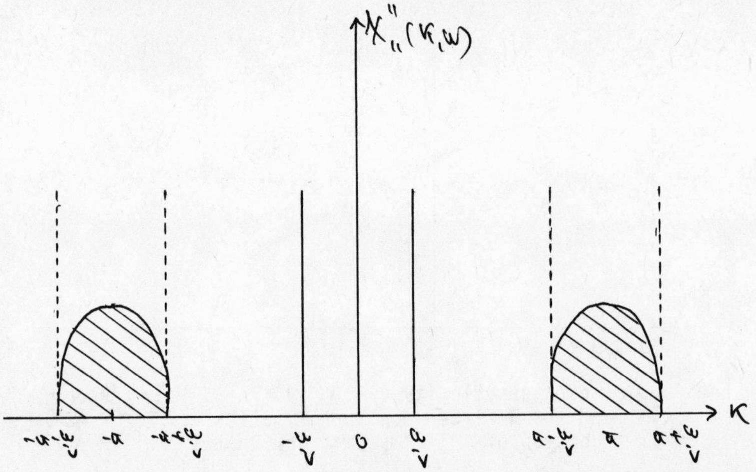

Fig.4.11 The ferromagnetic longitudinal response of the Heisenberg-Ising chain as a function of the wave-number k for fixed frequency ω (arbitrary units)

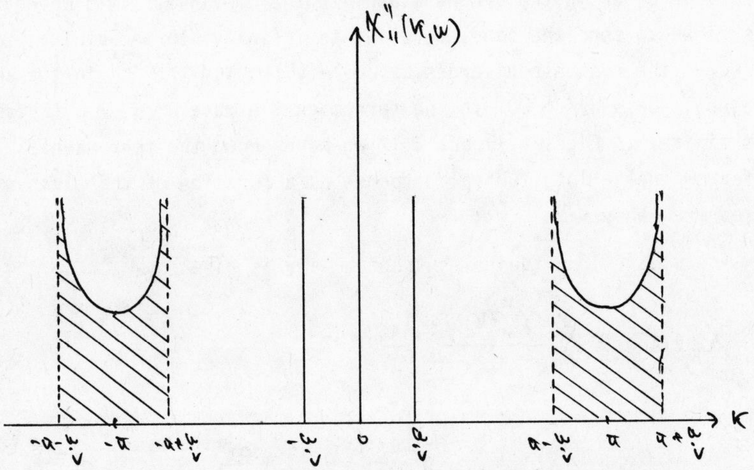

Fig.4.12 The antiferromagnetic longitudinal response of the Heisenberg-Ising chain as a function of the wave-number k for fixed frequency ω (arbitrary units)

In the long wavelength limit $k = 0$ the static susceptibility approaches $(1+2J''/\pi J^{\perp})/4\pi J^{\perp}$ showing that there is no long range order in the ground state, in conformity with the behaviour of the longitudinal correlations

(4.20). At the zone edges $\chi_{\parallel}(k)$ diverges as $|k\pm\pi|^{4J''/\pi J^{\perp}}$ in the antiferromagnetic case $J''/J^{\perp} < 0$, indicating that the system is soft towards a dimerisation with the period of the lattice (see, for instance ref.64). In the ferromagnetic case $J''/J^{\perp} > 0$ the static susceptibility vanishes for $k \to \pm\pi$. In Figs.4.13 and 4.14 we have shown the static ferromagnetic and antiferromagnetic longitudinal susceptibilities as functions of the wavenumber k.

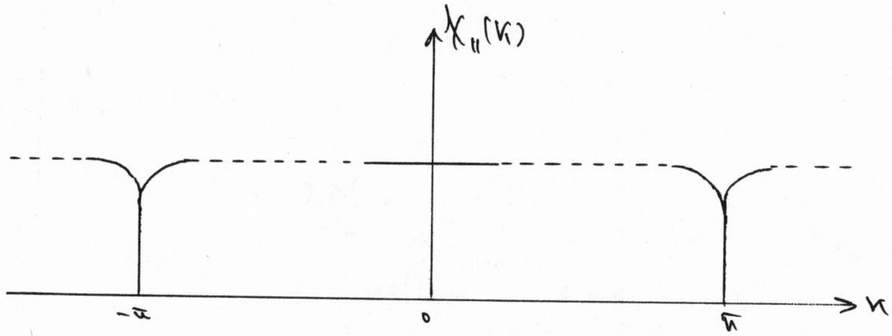

Fig.4.13 The static ferromagnetic longitudinal susceptibility for the Heisenberg-Ising chain as a function of the wavenumber k (arbitrary units)

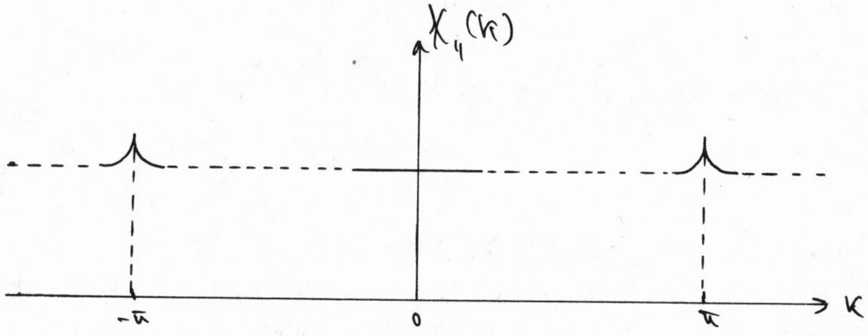

Fig.4.14 The static antiferromagnetic longitudinal susceptibility for the Heisenberg-Ising chain as a function of the wavenumber k (arbitrary units)

The transverse response (4.24) has band contributions both in the long wavelength limit and at the zone edges. Hence, there are no transverse spin waves in the Heisenberg-Ising chain. In the case of a ferromagnetic coupling in the basal plane, i.e., $J^{\perp} > 0$, the long wavelength response diverges as $(\omega^2 - v^2k^2)^{-3/4-J''/2\pi J^{\perp}}$ in the vicinity of the band edges $\pm vk$.

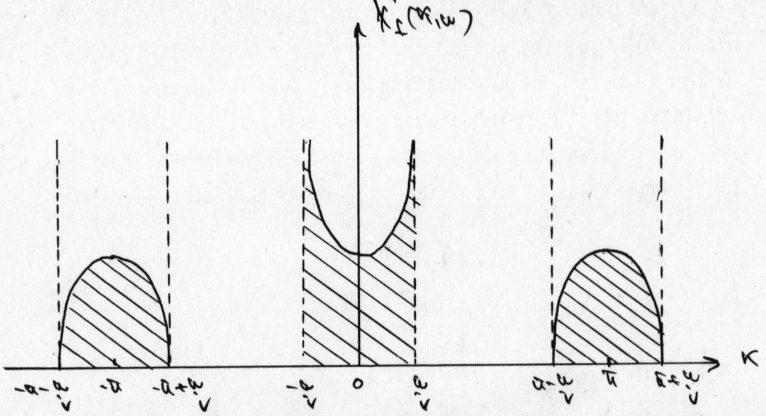

Fig.4.15 The ferromagnetic transverse response of the Heisenberg-Ising chain as a function of the wave-number k for fixed frequency ω (arbitrary units)

At the zone edge bands the response is finite. In Fig.4.15 we have shown the ferromagnetic transverse response as a function of the wavenumber k for fixed frequency ω.

The static transverse susceptibility is given by

$$\chi_\perp(k) = \text{cst.}\,|k|^{-\frac{3}{2}-\frac{J''}{\pi J^\perp}} + \text{cst.}\,|k-\pi|^{\frac{1}{2}+\frac{3J''}{\pi J^\perp}} + \text{cst.}\,|k+\pi|^{\frac{1}{2}+\frac{3J''}{\pi J^\perp}}$$

$$\text{for } |k| \ll \Lambda \ . \tag{4.27}$$

For k = 0 the static susceptibility $\chi_\perp(k)$ diverges with the exponent $-3/2-J''/\pi J^\perp$ indicating that the system is soft with respect to an overall rotation of the spins about the z axis. At the zone edges the static susceptibility vanishes as $k \to \pm\pi$ with exponent $\frac{1}{2} + \frac{3J''}{\pi J^\perp}$. The case of antiferromagnetic coupling in the basal plane, $J^\perp < 0$, is obtained by the transformation $S^\pm(x) \to (-1)^x S^\pm(x)$, which in Fourier space corresponds to $k \to k - \pi$ (see also Fig.4.4). Interchanging the band contributions in (4.24), we conclude that in the antiferromagnetic case $J^\perp < 0$ the transverse response diverges at the zone edges $k = \pm\pi$, showing that the system is susceptible to a dimerisation with period π. In Fig.4.16 we have shown the static transverse susceptibility in the case $J^\perp > 0$.

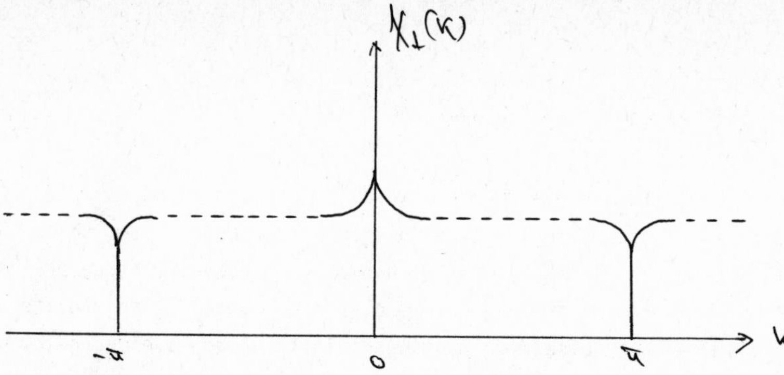

Fig.4.16 The static transverse susceptibility for the Heisenberg-Ising chain as a function of the wavenumber k in the case $J^{\perp} > 0$ (arbitrary units)

We finally remark that the interpretation of the longitudinal and transverse spectral functions for the Heisenberg-Ising model in terms of the underlying Luttinger model parallels the one given in Section 4.4 for the XY model.

CHAPTER V

HYDRODYNAMICS OF THE HEISENBERG PARAMAGNET

In the previous three chapters we discussed the ground state correlations and response properties of different variants of the spin half chain using a variety of quantum mechanical techniques. In the present context we turn our attention to a purely classical system at an elevated temperature, namely the Heisenberg paramagnet in the long wavelength-low frequency limit. This system has been considered by the author in ref. 65 within the framework of a non linear Langevin equation. Here we review and summarise the results obtained, and, furthermore, elaborate various points which were not treated in detail in ref. 65.

5.1 The Static Description of the Heisenberg Paramagnet

In the long wavelength-low frequency limit we can safely neglect quantum effects and thus base our discussion of the paramagnet on a semi-microscopic classical Heisenberg model,

$$H = -\sum_{ij} J_{ij} \bar{S}_i \bar{S}_j , \tag{5.1}$$

defined on a d-dimensional lattice with a lattice parameter a much larger than any microscopic de Broglie wavelength. Here $\bar{S}_i$ is a classical spin of length S associated with the site i and J_{ij} an exchange interaction of range λ_0. The static properties of the model are described by the free energy[66]

$$\bar{F} = -k_B T \log\left[\sum_{\{S\}} \exp(-H/(k_B T))\right],$$

where k_B is Boltzmann's constant and T the temperature. The sum $\sum_{\{S\}}$ is over all independent spin configurations. Performing a partial sum over spin modes with wavelengths in the interval $0 \leq \lambda \leq \lambda_0$, expanding the spin field, and incorporating the length condition, we can express the free energy in the continuum form

$$\bar{F} = -k_B T \log \int \prod_x d\bar{S}\, \delta(\bar{S}^2 - S^2) \exp\left(-\frac{J}{2}\int d^dx\, (\nabla \bar{S})^2\right),$$

where the integration $\int\prod_x d\bar{S} = \int\prod_x dSd\Omega_S$ is over the remaining spin modes with wavelengths in the interval $\lambda_0 \leq \lambda < \infty$. The effective exchange J is of order $J_{ij}a^{2-d}/k_BT$. The final step in constructing a long wavelength description of the paramagnet is achieved by representing the constraint $\bar{S}^2 = S^2$ by means of polynomial terms in the exponent. Scaling the spins in order to absorb the prefactor J, and keeping, for the present purposes, only up to and including quartic terms, we thus arrive at the well-known Ginzburg-Landau form for the free energy functional[36]

$$F = \frac{1}{2}\int d^dx\left[(\nabla\bar{S})^2 + r_0\bar{S}^2 + \frac{1}{2}u_0(\bar{S}^2)^2\right]. \tag{5.2}$$

The total free energy $\bar{F} = -k_BT \log \int\prod_x d\bar{S}\exp(-F)$ is obtained by averaging over the long wavelength spin configurations characterised by (5.2).

The thermodynamic equilibrium state corresponds to a minimum[65] of the free energy F. Since the state obviously is uniform, i.e., $\nabla\bar{S} = 0$, we obtain by differentiation with respect to $\bar{S}$ the condition $dF/d\bar{S} = [r_0+u_0\bar{S}^2]\,\bar{S} = 0$. In order to ensure that the equilibrium state is in the paramagnetic regime $\bar{S} = 0$ we choose the parameters r_0 and u_0 positive. Within a mean field description of critical phenomena[36] $r_0 = 0$ corresponds to the critical point, i.e., $r_0 = \text{cst.}(T-T_c)$, where T_c is the critical temperature. In the presence of a uniform magnetic field $\bar{h}$, corresponding to adding a term $-\bar{h}\cdot\bar{S}$ to the free energy density, the equilibrium condition reads $[r_0+u_0\bar{S}^2]\,\bar{S} = \bar{h}$, and we identify $1/r_0$ with the static susceptibility χ. In particular, $r_0 \propto T - T_c$ implies the well-known Curie-Weiss law[36] $\chi \propto 1/(T - T_c)$.

5.2 Conservation Laws - Slow Variables

In the long wavelength-low frequencylimit the dynamical evolution of a many body system is controlled by the conservation laws[21,67], which imply that a disturbance of a conserved density does not disappear locally but gives rise to a current and thus spreads out over the whole system. This observation is of crucial importance in constructing a closed hydrodynamical description of the long wavelength-low frequency properties of an interacting many body system. In most cases the conservation laws imply a natural time scale separation between "slow" conserved variables and "fast" microscopic variables.

For the purpose of setting up a dynamical description of the paramagnet our first task is therefore to identify the relevant conserved slow variables. The conservation laws are usually associated with the symmetry properties of the underlying Hamiltonian[68]. For the Heisenberg model (5.1) the invariance under rotations in spin space entails the conservation of the total spin $\simeq \int d^dx\bar{S}(x)$. Since also the total energy $\simeq \int d^dx(\nabla\bar{S})^2$ is conserved we have two slow variables for the Heisenberg paramagnet, namely the spin density $\bar{S}(x)$ and the energy density $\varepsilon(x) \propto (\nabla\bar{S})^2$, obeying the conservation laws or continuity equations,

$$\frac{dS^\alpha}{dt} + \bar{\nabla}\bar{j}^\alpha = 0 , \quad \alpha = x,y,z \tag{5.3a}$$

$$\frac{d\varepsilon}{dt} + \bar{\nabla}\cdot\bar{j} = 0 . \tag{5.3b}$$

Here the d-dimensional vectors $\bar{j}^\alpha$ and $\bar{j}$ are the associated spin and energy currents.

We have deliberately omitted two additional slow variables, namely the angular momentum density ℓ associated with the invariance of the Hamiltonian under rotations in d-dimensional space, and the momentum density p related to the invariance under space translations. Since the spins are confined to a lattice, rotations in space and spin space are completely independent, and the "orbital" angular momentum density ℓ, therefore, does not enter in the dynamical description of the paramagnet. In Chapter VI we show that the momentum density in the long wavelength limit has the form $p \propto (S^x\nabla S^y - S^y\nabla S^x)/(1+S^z)$, and thus depends on the choice of spin coordinate system. The momentum density, in other words, is not invariant under rotations in spin space and, hence, cannot couple to either the rotationally invariant energy density or to the spin density.

Since the variables $\bar{S}$ and ε enter the dynamical description on an equal footing it turns out to be convenient to include the energy density in the free energy functional (5.2), although it here will have a redundant character as far as the static properties are concerned. Following Halperin et al.[69] (see also the renct review ref. 70) we introduce the free energy

$$F = \frac{1}{2}\int d^dx\left[r_0\bar{S}^2 + (\nabla\bar{S})^2 + \frac{1}{2}u_0(\bar{S}^2)^2 + c_0\varepsilon^2 + g_0\varepsilon\bar{S}^2\right], \tag{5.4}$$

where we have only kept leading terms in ε and allowed for the time reversal invariant coupling $g_0 \varepsilon \bar{S}^2$. Minimising F with respect to variations in $\bar{S}$ and ε in order to determine the uniform equilibrium state we obtain $[r_0 + \bar{u}_0 \bar{S}^2 + g_0 \varepsilon]\,\bar{S} = 0$ and $2c_0\varepsilon + g_0\bar{S}^2 = 0$ or, eliminating ε, $r_0 + (\bar{u}_0 - g_0^2/2c_0)\bar{S}^2 = 0$ which shows that the energy density can be absorbed by introducing $u_0 = \bar{u}_0 - g_0^2/2c_0$. The same result is, of course, achieved by integrating out the energy density directly, and we conclude that the static properties of (5.4) are identical to those of (5.2) with $\bar{u}_0 = u_0 + g_0^2/2c_0$.

5.3 Hydrodynamical Equations for the Paramagnet

A closed macroscopic description of the long wavelength-low frequency properties of an interacting many body system is based on three ingredients, namely the conservation laws or continuity equations for the slow variables, the transport equations or constitutive relations for the associated currents, and the assumption of local thermodynamic equilibrium[21,67,71].

For the paramagnet the relevant slow variables are, as discussed above, the spin density S^α and the energy density ε satisfying the conservation laws (5.3a-b), i.e., $dS^\alpha/dt + \bar{\nabla}\,\bar{j}^\alpha = 0$ and $d\varepsilon/dt + \bar{\nabla}\cdot\bar{j} = 0$. The transport equations are obtained by expanding the spin current $\bar{j}^\alpha$ and the energy current $\bar{j}$ in powers of gradients ∇ and in powers of the field amplitudes ε and S^α, allowing for all terms compatible with the symmetry of the problem.

To linear order in the amplitudes we have the transport equations,

$$\bar{j}^\alpha = -D_0 \bar{\nabla} S^\alpha , \tag{5.5a}$$

$$\bar{j} = -E_0 \bar{\nabla} \varepsilon , \tag{5.5b}$$

which by insertion in the conservation laws (5.3a-b) yield the hydrodynamical equations governing spin and energy diffusion in the paramagnet

$$\frac{dS^\alpha}{dt} = D_0 \nabla^2 S^\alpha \tag{5.6a}$$

$$\frac{d\varepsilon}{dt} = E_0 \nabla^2 \varepsilon . \tag{5.6b}$$

Here D_0 and E_0 are the spin and energy diffusion coefficents, respectively. We note that to linear order the diffusion equations are purely irreversible, furthermore, the spin and energy modes are decoupled.

Beyond linear order the transport equations for $\bar{J}^{\alpha}$ and $\bar{J}$ include both reversible and irreversible terms. The reversible contributions arise from the microscopic spin dynamics of the paramagnet. Retaining terms to second order in the amplitudes we obtain the constitutive relations

$$\bar{J}^{\alpha} = -D_0\bar{\nabla}S^{\alpha} - \lambda_0\sum_{\beta\gamma}\varepsilon^{\alpha\beta\gamma}S^{\beta}\bar{\nabla}S^{\gamma} - A_0\varepsilon\bar{\nabla}S^{\alpha} - B_0S^{\alpha}\bar{\nabla}\varepsilon \tag{5.7a}$$

$$\bar{J} = -E_0\bar{\nabla}\varepsilon - G_0\bar{\nabla}S^2, \quad S^2 = \sum_{\alpha}(S^{\alpha})^2 \tag{5.7b}$$

which, inserted in the conservations laws, lead to the non linear hydrodynamical equations

$$\frac{dS^{\alpha}}{dt} = \lambda_0\sum_{\beta\gamma}\varepsilon^{\alpha\beta\gamma}S^{\beta}\nabla^2S^{\gamma} + D_0\nabla^2S^{\alpha} + A_0\varepsilon\nabla^2S^{\alpha} + B_0S^{\alpha}\nabla^2\varepsilon + (A_0+B_0)\nabla S^{\alpha}\nabla\varepsilon$$

$$\frac{d\varepsilon}{dt} = E_0\nabla^2\varepsilon + G_0\nabla^2S^2 \tag{5.8b}$$

To second order the only reversible contribution is the precessional term $\lambda_0\bar{S}\times\nabla^2\bar{S}$ due to the motion of the spin density in the local magnetic field $\sim\nabla^2\bar{S}$ generated by the neighbouring spins via the exchange interaction. We also notice that the non linear terms give rise to a coupling between the energy and spin modes.

In the derivation of the mode coupling equations (5.8a-b) for the paramagnet we have implicitly assumed that the densities S^{α} and ε and the currents $\bar{J}^{\alpha}$ and $\bar{J}$ are non-fluctuating space and time dependent mean values evaluated in an unspecified non-equilibrium ensemble. In order to obtain a dynamical description compatible with the equilibrium state characterised by the distribution function exp(-F) we must consider the thermal fluctuations, i.e., the stochastic nature of the problem.

5.4 Markov Process - Fokker Planck Equation

The hydrodynamical description of the paramagnet advanced in the preceeding section essentially implies a complete separation of time scales. In "coarse grained" phase space the system is characterised by the macroscopic slow variables S^α and $\mathcal{E}$ obeying the conservation laws (5.3a-b). The microscopic fast variables do not enter directly in the hydrodynamical description but give rise to random fluctuations of S^α and $\mathcal{E}$, the well-known Brownian motion[66,71]. The variables S^α and $\mathcal{E}$ are thus of stochastic nature and the dynamical state of the system is described by a time dependent normalised distribution function $P(S^\alpha \mathcal{E} t)$, $\int \prod dS^\alpha \, d\mathcal{E} \, P(S^\alpha, \mathcal{E}, t) = 1$. In the thermodynamic equilibrium state the distribution is time independent and given by[66]

$$P_0(S^\alpha, \mathcal{E}) = \exp(-F(S^\alpha, \mathcal{E}))/Z, \tag{5.10}$$

where $F(S^\alpha, \mathcal{E})$ is the time reversal invariant free energy functional (5.4) and the normalisation factor

$$Z = \exp(-\bar{F}/k_B T) = \int \prod dS^\alpha d\mathcal{E} \, \exp(-F(S^\alpha, \mathcal{E}))$$

the usual partition function.

For a continuous stochastic process without memory effects, i.e., a Markov process, characterising the strict separation of time scales, the distribution function satisfies a linear partial differential equation, the Fokker Planck equation. Before we derive that equation in the present context let us briefly review the theory of a time dependent Markov process for a single stochastic variable, following ref. 72.

For a stationary Markov process the probability $P(x't')$ of finding $x = x'$ at $t = t'$ is given by

$$P(x', t') = \int W(x't', x''t'') \, P(x'', t'') \, dx'',$$

where $W(x't'; x''t'') = W(x'x''; t'-t'')$ is the conditional transition probability of realising $x = x'$ at $t = t'$ provided $x = x''$ at $t = t''$. In Fig. 5.1 we have shown a Markov process, the dashed line indicates the absence of memory effects. The probability $P(x', t')$ only depends on the preceeding $P(x'', t'')$ and the transition probability $W(x't'; x''t'')$.

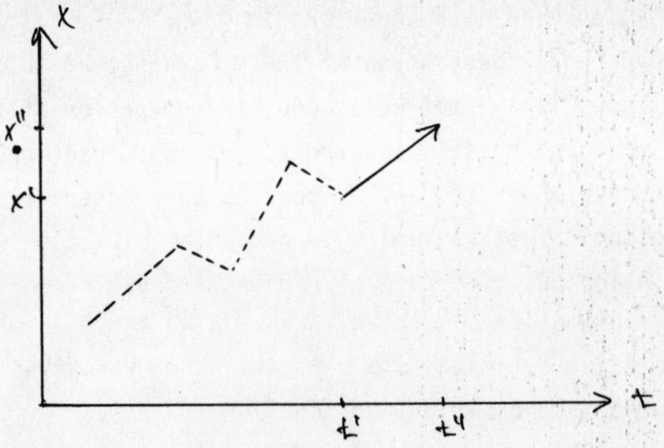

Fig. 5.1 Markov process, the absence of memory effects is indicated by the dashed line

The transition probability $W(x't', x"t")$, depicted in Fig. 5.2, is conveniently characterised by its moments M_n,

$$M_n(t't", x") = \int (x' - x")^n W(x't', x"t") \, dx'$$

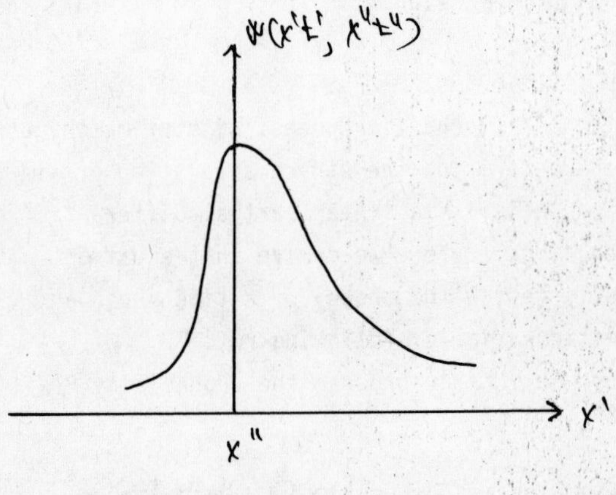

Fig. 5.2 The conditional transition probability $W(x't', x"t")$

The first moment $M_1(t't'';x'')$ describes the asymmetry of W, the second moment $M_2(t't'';x'')$ is a measure of the width, etc. The moments M_n completely specify the distribution W as can be seen by introducing the characteristic function

$$\Theta(t't'';kx'') = \int \exp(ik(x'-x''))\, W(x't';x''t'')\, dx' .$$

Expanding the exponential we have

$$\Theta(t't'';kx'') = \sum_{n=0}^{\infty} \frac{(ik)^n}{n!} M_n(t't'',x'')$$

which by insertion in

$$W(x't';x''t'') = \int \exp(-ik(x'-x''))\, \Theta(t't'';kx'')\frac{dk}{2\pi}$$

yields

$$W(x't';x''t'') = \sum_{n=0}^{\infty} \frac{1}{n!}\left(-\frac{d}{dx}\right)^n \delta(x'-x'')\, M_n(t't'';x'') .$$

Substituting W in $P = \int WPdx''$ and noting that $M_0 = \int W(x't';x''t'')dx' = 1$ we obtain

$$P(x't') - P(x't) = \sum_{n=1}^{\infty} \frac{1}{n!}\left(-\frac{d}{dx'}\right)^n [M_n(t't;x')\, P(x't)] .$$

Dividing by t'-t and performing the limit $t'-t \to 0^+$ we arrive at a partial differential equation for the distribution function P(xt)

$$\frac{dP(xt)}{dt} = \sum_{n=1}^{\infty} \frac{1}{n!}\left(-\frac{d}{dx}\right)^n (K_n(x)P(xt)) ,$$

where the intensity coefficients $K_n(x)$ are defined by

$$K_n(x) = \lim_{t'-t''\to 0^+} \frac{M_n(t't'',x)}{t'-t''} .$$

In the limit $t'-t'' \to 0$ the distribution W approaches $\delta(x'-x'')$ and the existence of the intensity coefficients imply that the moments of W vanish at least as fast as t'-t'', i.e., $M_n(t't'';x) = (t'-t'')K_n(x) + O((t'-t'')^2)$.

In the case $K_n = 0$ for $n \geqslant 3$ we have a continuous stationary Markov process and we obtain the Fokker-Planck equation

$$\frac{dP(x,t)}{dt} + \frac{d}{dx}[K_1(x)P(x,t)] - \frac{1}{2}\frac{d^2}{dx^2}[K_2(x)P(x,t)] = 0 \qquad (5.10)$$

for the distribution function $P(x,t)$.

One notices that (5.10) has the form of a continuity equation, $dP/dt + dJ/dx = 0$, expressing the conservation of the total probability, $\int P(x)dx = 1$. The probability current $J(x,t)$ is composed of two parts, a drift term $K_1(x)P(x,t)$ and a diffusive term $-\frac{1}{2}\frac{d}{dx}(K_2(x)P(x,t))$. The drift term characterises the coherent flow in phase space $\{x\}$, whereas the diffusive term is a measure of the loss of probability due to the stochastic background.

Let us establish under which conditions the time dependent distribution $P(xt)$ can approach the stationary equilibrium distribution P_0 specified by the free energy function $F(x)$, $P_0 \propto \exp(-F)$. For the equilibrium distribution we have $dJ_0/dx = 0$, i.e.,

$$\frac{d}{dx}[K_1(x)\exp(-F)] - \frac{1}{2}\frac{d^2}{dx^2}[K_2(x)\exp(-F)] = 0,$$

which is satisfied provided the potential conditions

$$K_2(x)\frac{dF(x)}{dx} = \frac{d}{dx}K_2(x) - 2I_1(x) \qquad (5.11a)$$

$$R_1(x)\frac{dF(x)}{dx} = \frac{d}{dx}R_1(x) \qquad (5.11b)$$

hold. Here we have set $K_1 = R_1 + I_1$, where R_1 and I_1 denote the reversible and irreversible part of the coherent drift, respectively. The first condition expresses the balance between the irreversible part of the coherent flow and the background. The second relation ensures that the reversible coherent drift does not perturb the time reversal invariant equilibrium distribution.

Instead of discussing the time development of the distribution function $P(x,t)$ according to the Fokker-Planck equation (5.10), it is conceptually simpler and more practical from the point of view of a calculation to consider an equivalent equation of motion[72] for the random variable itself (for the derivation we refer to ref. 72),

$$\frac{dx}{dt} = K_1(x) - \frac{1}{2}\frac{dG(x)}{dx}G(x) + G(x)f(t).$$

Here $G(x)^2 = K_2(x)$ and $f(t)$ is a delta function correlated white noise, i.e., $\langle f(t)\rangle = 0$, $\langle f(t)f(t')\rangle = \delta(t-t')$. Eliminating the irreversible part of the drift term by means of the potential condition (5.11a) we obtain

$$\frac{dx}{dt} = R_1(x) - \frac{1}{2}K_2(x)\frac{dF}{dx} + \frac{1}{4}\frac{d}{dx}K_2(x) + \sqrt{K_2(x)}\,f(t). \quad (5.12)$$

$$\langle f(t)\rangle = 0 \quad \langle f(t)f(t')\rangle = \delta(t-t')$$

This is the most general Langevin equation describing the random motion of a single variable whose distribution $P(xt)$ satisfies the Fokker-Planck equation (5.10) and approaches the equilibrium distribution $\exp(-F)$. The direct interpretation of (5.12) is straightforward (see, for instance, Graham and Haken[73,74]). The irreversible drift term $-\frac{1}{2}K_2(x)\frac{dF}{dx}$ drives the variable x towards x_0 for which F has a minimum, i.e., $(dF/dx)_{x_0} = 0$. At $x = x_0$ the potential condition (5.11b) implies $dR_1/dx = 0$ and the reversible drift $R_1(x)$, therefore, does not alter the equilibrium distribution. The Brownian motion is imparted by the noise term $(K_2(x))^{1/2}f(t)$, where $f(t)$ has a flat spectrum. Since the noise depends on x it contains a spurious drift which is compensated by the irreversible drift term $\frac{1}{4}dK_2/dx$.

The Fokker-Planck and Langevin equations (5.10) and (5.12) are easily generalised to the case of several stochastic variables $x_1, x_2, \ldots$ Referring to refs. 72, 73, and 74 for details, we give the corresponding equations below. The Fokker-Planck equation for the distribution function $P(\{x_n\}, t)$ takes the form

$$\frac{dP}{dt} + \sum_n \frac{d}{dx_n}[(R_n + I_n)P] - \frac{1}{2}\sum_{nm}\frac{d^2}{dx_n dx_m}[K_{nm}P] = 0, \quad (5.13)$$

where R_n and I_n are the reversible and irreversible drifts, respectively; K_{nm} is the diffusion matrix. In order that P approaches the time reversal invariant equilibrium distribution $P_0(\{x_n\}) = \exp(-F(\{x_n\}))$, where F is the free energy, the following potential conditions must be satisfied,

$$\sum_n K_{pn} \frac{dF}{dx_n} = \sum_n \frac{d}{dx_n} K_{pn} - 2I_p \tag{5.14a}$$

$$\sum_n \frac{d}{dx_n} R_n = \sum_n R_n \frac{dF}{dx_n} . \tag{5.14b}$$

The Langevin equation for x_n corresponding to (5.13) has the form

$$\frac{dx_p}{dt} = R_p + I_p - \frac{1}{2}\sum_{nm} G_{mn} \frac{dG_{pn}}{dx_m} + \sum_n G_{pn} \xi_n(t) \tag{5.15}$$

$$\sum_n G_{pn} G_{mn} = K_{pm} , \quad \langle \xi_n(t) \rangle = 0 , \quad \langle \xi_n(t) \xi_m(t') \rangle = \delta_{nm} \delta(t-t') .$$

Notice that G_{nm} can be chosen arbitrarily, subject to the condition $\sum_n G_{pn} G_{nm} = K_{pm}$. In the case where G_{pn} is independent of x_p the compensating drift vanishes and we obtain the standard Langevin equation

$$\frac{dx_p}{dt} = R_p - \frac{1}{2} \sum_n K_{pn} \frac{dF}{dx_n} + f_p(t) ,$$

where we have used the potential condition (5.14a) in order to eliminate I_p and, furthermore, introduced the noise $f_p(t)$ satisfying the Einstein relation

$$\langle f_n(t) f_m(t') \rangle = K_{nm} \, \delta(t-t') .$$

5.5 Non Linear Langevin Equations for the Paramagnet

In the previous section we summarised the general theory of a continuous Markov process. Here we turn to the construction of the Fokker-Planck and Langevin equations for the paramagnet. We have now two random functions, namely the spin density $S^\alpha(x)$ and the energy density $\varepsilon(x)$. The dynamical state of the system is characterised by the distribution functional $P(S^\alpha, \varepsilon, t)$. By a straightforward continuum generalisation of (5.13) we obtain the Fokker-Planck equation

$$\frac{dP}{dt} + \int d^d x \left[\sum_\alpha \frac{d}{dS^\alpha} J^\alpha + \frac{d}{d\varepsilon} J^0 \right] = 0 , \tag{5.16}$$

where the probability currents J^{α} and J^{0} in terms of the drifts R and I and the diffusion matrix K are given by

$$J^{\alpha} = (R^{\alpha} + I^{\alpha})P - \frac{1}{2}\int d^dx \left[\sum_{\beta} \frac{d}{ds^{\beta}}(K^{\alpha\beta}P) + \frac{d}{d\varepsilon}(K^{\alpha 0}P) \right] \tag{5.17a}$$

$$J^{0} = (R^{0} + I^{0})P - \frac{1}{2}\int d^dx \left[\sum_{\beta} \frac{d}{ds^{\beta}}(K^{0\beta}P) + \frac{d}{d\varepsilon}(K^{00}P) \right] \tag{5.17b}$$

Similarly, the potential conditions (5.14a) and (5.14b), ensuring the approach to the equilibrium distribution exp(-F), take the form

$$I^{0} = \frac{1}{2}\int d^dx \left[\sum_{\beta} \frac{dK^{0\beta}}{ds^{\beta}} + \frac{dK^{00}}{d\varepsilon} \right] - \frac{1}{2}\int d^dx \left[\sum_{\beta} K^{0\beta}\frac{dF}{ds^{\beta}} + K^{00}\frac{dF}{d\varepsilon} \right] \tag{5.18a}$$

$$I^{\alpha} = \frac{1}{2}\int d^dx \left[\sum_{\beta} \frac{dK^{\alpha\beta}}{ds^{\beta}} + \frac{dK^{\alpha 0}}{d\varepsilon} \right] - \frac{1}{2}\int d^dx \left[\sum_{\beta} K^{\alpha\beta}\frac{dF}{ds^{\beta}} + K^{\alpha 0}\frac{dF}{d\varepsilon} \right] \tag{5.18b}$$

$$\int d^dx \left[\sum_{\beta} \frac{dR^{\beta}}{ds^{\beta}} + \frac{dR^{0}}{d\varepsilon} \right] = \int d^dx \left[\sum_{\beta} R^{\beta}\frac{dF}{ds^{\beta}} + R^{0}\frac{dF}{d\varepsilon} \right]. \tag{5.19}$$

In analogy with (5.15) the corresponding Langevin equations are

$$\begin{aligned} \frac{ds^{\alpha}}{dt} = {} & R^{\alpha} + I^{\alpha} - \frac{1}{2}\int d^dx\, d^dy \Big[\sum_{\beta\gamma} \frac{dG^{\alpha\beta}}{ds^{\gamma}} G^{\gamma\beta} + \sum_{\beta} \frac{dG^{\alpha\beta}}{d\varepsilon} G^{0\beta} \\ & + \sum_{\beta} \frac{dG^{\alpha 0}}{ds^{\beta}} G^{\beta 0} + \frac{dG^{\alpha 0}}{d\varepsilon} G^{00} \Big] + \int d^dx \Big[\sum_{\beta} G^{\alpha\beta} \xi^{\beta}(t) + G^{\alpha 0}\xi^{0}(t) \Big] \end{aligned} \tag{5.20a}$$

$$\begin{aligned} \frac{d\varepsilon}{dt} = {} & R^{0} + I^{0} - \frac{1}{2}\int d^dx\, d^dy \Big[\sum_{\alpha\beta} \frac{dG^{0\alpha}}{ds^{\beta}} G^{\beta\alpha} + \sum_{\alpha} \frac{dG^{0\alpha}}{d\varepsilon} G^{0\alpha} \\ & + \sum_{\alpha} \frac{dG^{00}}{ds^{\alpha}} G^{\alpha 0} + \frac{dG^{00}}{d\varepsilon} G^{00} \Big] + \int d^dx \Big[\sum_{\beta} G^{0\beta} \xi^{\beta}(t) + G^{00}\xi^{0}(t) \Big] \end{aligned} \tag{5.20b}$$

for $\langle \xi^{\alpha}(t)\xi^{\beta}(t')\rangle = \delta^{\alpha\beta}\delta(t-t')$, $\langle \xi^{0}(t)\xi^{0}(t')\rangle = \delta(t-t')$
and $\langle \xi^{\alpha}(t)\xi^{0}(t')\rangle = 0$,

where the arbitrary matrix G is constrained by the conditions

$$\int d^dx \left[\sum_{\gamma} G^{\alpha\gamma}G^{\beta\gamma} + G^{\alpha 0}G^{\beta 0} \right] = K^{\alpha\beta} \tag{5.21a}$$

$$\int d^dx \left[\sum_{\gamma} G^{0\gamma}G^{\beta\gamma} + G^{00}G^{\beta 0} \right] = K^{0\beta} \tag{5.21b}$$

$$\int d^dx \left[\sum_{\gamma} G^{\alpha\gamma}G^{0\gamma} + G^{\alpha 0}G^{00} \right] = K^{\alpha 0} \tag{5.21c}$$

$$\int d^dx \left[\sum_{\gamma} G^{0\gamma}G^{0\gamma} + G^{00}G^{00} \right] = K^{00} \tag{5.21d}$$

The irreversible drifts I^0 and I^α are given by the potential conditions (5.18a) and (5.18b) from our choice of free energy F and diffusion matrix K. The reversible drifts R^0 and R^α are determined by the underlying microscopic Poisson bracket relations for the paramagnet[36,75]. We must, furthermore, build in the conservation laws (5.3a-b) by an appropriate choice of the coefficients K. Expanding K in powers of gradients and fields we obtain to lowest non-trivial order

$$K^{\alpha\beta}(xx') = \delta^{\alpha\beta}[-2\Gamma_0\nabla^2\delta^d(x-x') + 2\eta_0\nabla\nabla'\varepsilon(x)\,\delta^d(x-x')] \qquad (5.22a)$$

$$K^{0\beta}(xx') = K^{\alpha 0}(xx') = 0 \qquad (5.22b)$$

$$K^{00}(xx') = -2\gamma_0\nabla^2\delta^d(x-x'). \qquad (5.22c)$$

By insertion in (5.18a-b) the irreversible drift terms are

$$I^0(x) = \gamma_0\nabla^2\frac{dF}{d\varepsilon(x)} \qquad (5.23a)$$

$$I^\alpha(x) = \Gamma_0\nabla^2\frac{dF}{dS^\alpha(x)} + \eta_0\nabla(\varepsilon(x)\nabla\frac{dF}{dS^\alpha(x)}). \qquad (5.23b)$$

Since the scalar $\varepsilon(x)$ is invariant under time reversal, a reversible mode coupling term involves at least three spins, discarding such terms we set $R^0 = 0$. The potential condition (5.19) then takes the form

$$\int d^dx\,[\sum_\alpha\frac{dR^\alpha}{dS^\alpha} - \sum_\alpha R^\alpha\frac{dF}{dS^\alpha}] = 0.$$

Choosing

$$R^\alpha(x) = -\lambda_0\sum_{\beta\gamma}\varepsilon^{\alpha\beta\gamma}S^\beta(x)\frac{dF}{dS^\gamma(x)} \qquad (5.24)$$

the last term vanishes by symmetry and we have $\sum_\alpha dR^\alpha(x)/dS^\alpha(x) = 0$ which is Liouville's theorem in phase space[36,66]. In order to complete the construction of the Langevin equations for the paramagnet we must choose the G's such as to satisfy (5.21a-d). Setting $G^{0\alpha} = G^{\alpha 0} = 0$ and $G^{\alpha\beta}(xx') = \delta^{\alpha\beta}G(xx')$, i.e.

$$\int G^{00}(xx'')G^{00}(x'x'')d^dx'' = -2\gamma_0\nabla^2\delta^d(x-x') \tag{5.25a}$$

$$\int G(xx'')G(x'x'')d^dx'' = -2\Gamma_0\nabla^2\delta^d(x-x') + 2\eta_0\nabla\nabla'\varepsilon(x)\delta^d(x-x') \tag{5.25b}$$

and piecing things together we obtain two non linear Langevin equations for the spin density $S^\alpha(x)$ and the energy density $\varepsilon(x)$,

$$\frac{dS^\alpha}{dt} = -\lambda_0\sum_{\beta\gamma}\varepsilon^{\alpha\beta\gamma}S^\beta\frac{dF}{dS^\gamma} + \nabla\left((\Gamma_0+\eta_0\varepsilon)\nabla\frac{dF}{dS^\alpha}\right) + f^\alpha \tag{5.26a}$$

$$\langle f^\alpha(xt)f^\beta(x't')\rangle = -2\delta^{\alpha\beta}\delta(t-t')\nabla((\Gamma_0+\eta_0\varepsilon)\nabla\delta^d(x-x'))$$

$$\frac{d\varepsilon}{dt} = \gamma_0\nabla^2\frac{dF}{d\varepsilon} + f \tag{5.26b}$$

$$\langle f(xt)f(x't')\rangle = -2\delta(t-t')\gamma_0\nabla^2\delta^d(x-x').$$

In the derivation of 5.26a) and (5.26b) we have only kept leading terms in ∇ and the leading non linear contributions. The reversible mode coupling term $-\lambda_0 S\times dF/dS$ describes the precession of the spin density in the local field $-\frac{dF}{dS}$. The irreversible damping term $\nabla((\Gamma_0+\eta_0\varepsilon)\nabla\, dF/dS)$ drives the system towards the minimum of F, i.e. the equilibrium state. The form of the noise term ensures detailed balance in the equilibrium state, that is the fluctuation-dissipation theorem. We notice that in contrast to the usual Langevin equation description, the damping term $\nabla((\Gamma_0+\eta_0\varepsilon)\nabla\, dF/dS)$ depends explicitly on the fluctuating energy density; this effect is, however, exactly balanced by the form of the spin noise correlations.

From the free energy functional (5.4) describing the equilibrium properties of the paramagnet we derive

$$\frac{dF}{dS^\alpha} = (r_0 - \nabla^2 + u_0S^2 + g_0\varepsilon)S^\alpha$$

$$\frac{dF}{d\varepsilon} = c_0\varepsilon + \frac{1}{2}g_0S^2$$

which by insertion in (5.26a-b) yield

$$\frac{dS^\alpha}{dt} = \lambda_0 \sum_{\beta\gamma} \varepsilon^{\alpha\beta\gamma} S^\beta \nabla S^\gamma + \nabla((\Gamma_0 + \eta_0 \varepsilon)\nabla(r_0 - \nabla^2 + u_0 S^2 + g_0 \varepsilon)S^\alpha) + f^\alpha$$

$$\langle f^\alpha(xt) f^\beta(x't')\rangle = -2\delta^{\alpha\beta}\delta(t-t')\nabla((\Gamma_0 + \eta_0\varepsilon)\nabla\delta^d(x-x')) \tag{5.27a}$$

$$\frac{d\varepsilon}{dt} = \gamma_0 \nabla^2(c_0\varepsilon + \tfrac{1}{2} g_0 S^2) + f$$

$$\langle f(xt) f(x't')\rangle = -2\delta(t-t')\gamma_0\nabla^2\delta^d(x-x') . \tag{5.27b}$$

We remark that the non linear Langevin equations (5.27a-b) have the same structure, barring the noise terms, as the hydrodynamical equations derived in Section 5.3 provided we make the identifications: $A_0 = \eta_0 r_0 + \Gamma_0 g_0$, $B_0 = \Gamma_0 g_0$, $G_0 = \frac{1}{2}\gamma_0 g_0$, $D_0 = \Gamma_0 r_0$, and $E_0 = \gamma_0 c_0$.

5.6 Classical Perturbation Theory for the Paramagnet

Here we begin a more detailed discussion of the non linear Langevin equations derived in the previous section. In contrast to the hydrodynamical equations (5.8a-b) for the time and space dependent mean values of S^α and ε, the Langevin equations describe the interaction between the fluctuating spin and energy modes and, furthermore, ensure that the system settles to the equilibrium state characterised by the free energy (5.4). We wish in particular to investigate the influence of the non linear mode coupling terms on the long wavelength-low frequency dynamics. This analysis is the content of ref. 65.

In the linear small amplitude regime the Langevin equations reduce to two driven diffusion equations for the spin and energy fluctuations,

$$\frac{dS^\alpha}{dt} = \Gamma_0 (r_0 - \nabla^2)\nabla^2 S^\alpha + f^\alpha \tag{5.28a}$$

$$\langle f^\alpha(xt) f^\beta(x't')\rangle = -2\delta^{\alpha\beta}\Gamma_0\delta(t-t')\nabla^2\delta^d(x-x')$$

$$\frac{d\varepsilon}{dt} = \gamma_0 c_0 \nabla^2\varepsilon + f \tag{5.28b}$$

$$\langle f(xt) f(x't')\rangle = -2\gamma_0\delta(t-t')\nabla^2\delta^d(x-x')$$

In Fourier space, $S^{\alpha}_{k\omega} = \int \exp(i\omega t - ikx) S^{\alpha}(xt) d^d x dt$, etc., we obtain the solutions

$$S^{\alpha}_{k\omega} = \frac{f^{\alpha}_{k\omega}}{-i\omega + \Gamma_0 (r_0 + k^2) k^2}$$

$$\langle f^{\alpha}_{k\omega} f^{\beta}_{p\nu} \rangle = 2\delta^{\alpha\beta} \Gamma_0 k^2 (2\pi)^{1+d} \delta(\omega+\nu) \delta^d(k+p)$$

$$\varepsilon_{k\omega} = \frac{f_{k\omega}}{-i\omega + \gamma_0 c_0 k^2}$$

$$\langle f_{k\omega} f_{p\nu} \rangle = 2\gamma_0 k^2 (2\pi)^{1+d} \delta(\omega+\nu) \delta^d(k+p)$$

or, in time space

$$S^{\alpha}_{k}(t) = -i \int_{-\infty}^{t} \exp(-\Gamma_0 (r_0 + k^2) k^2 (t-t')) f^{\alpha}_{k}(t') dt'$$

$$\langle f^{\alpha}_{k}(t) f^{\beta}_{p}(t') \rangle = 2\delta^{\alpha\beta} \Gamma_0 k^2 (2\pi)^d \delta^d(k+p) \delta(t-t')$$

$$\varepsilon_{k}(t) = -i \int_{-\infty}^{t} \exp(-\gamma_0 c_0 k^2 (t-t')) f_{k}(t') dt'$$

$$\langle f_{k}(t) f_{p}(t') \rangle = 2\gamma_0 k^2 (2\pi)^d \delta^d(k+p) \delta(t-t')$$

which show that the spin and energy correlations decay exponentially with life times $\tau_s = 1/\Gamma_0 (r_0 + k^2) k^2$ and $\tau_\varepsilon = 1/\gamma_0 c_0 k^2$. In the long wavelength limit $k = 0$ τ_s and τ_ε become infinite, a characteristic feature of hydrodynamical modes controlled by conservation laws[21,67].

Adding the non random forces F^{α} and F to (5.28a) and (5.28b) the response of the mean spin and energy $\langle S^{\alpha} \rangle$ and $\langle \varepsilon \rangle$ is in Fourier space given by

$$\delta \langle S^{\alpha}_{k\omega} \rangle = G_0(k\omega) F^{\alpha}_{k\omega}$$

$$\delta \langle \varepsilon_{k\omega} \rangle = D_0(k\omega) F_{k\omega} \,,$$

where we have introduced the response functions, or susceptibilities

$$G_0(k,\omega) = \frac{1}{-i\omega + \Gamma_0 (r_0 + k^2) k^2} \,, \tag{5.29a}$$

$$F_0(k,\omega) = \frac{1}{-i\omega + \gamma_0 C_0 k^2} \,. \tag{5.29b}$$

We notice that the diffusive behaviour of the spin and energy fluctuations is here characterised by a purely imaginary pole of the corresponding response function. In Fig. 5.3 we have shown the exponential decay of $S^{\alpha}_{k}(t)$ and in Fig. 5.4 the corresponding pole structure of $G_0(k,\omega)$.

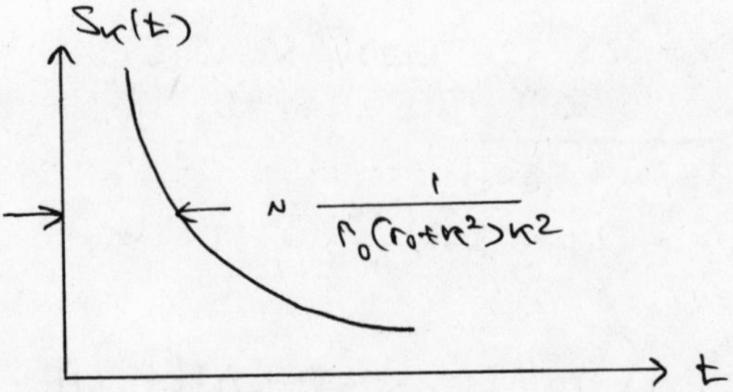

Fig. 5.3 The exponential decay of a spin fluctuation

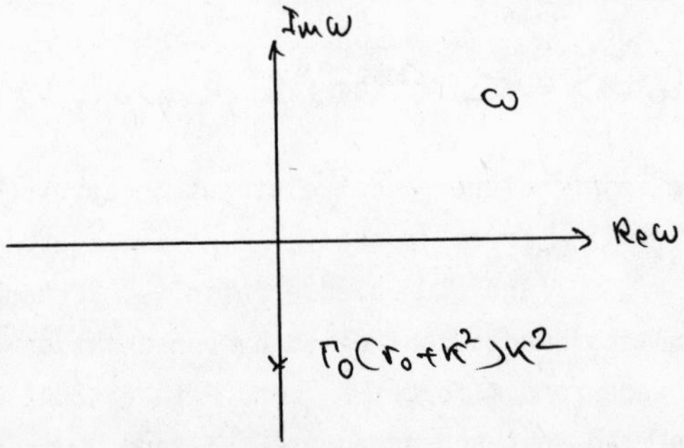

Fig. 5.4 The pole structure of $G_0(k\omega)$ in the case of diffusion

The fluctuation spectra of $S^{\alpha}_{k\omega}$ and $\varepsilon_{k\omega}$ are characterised by the correlation functions

$$C_0(k,\omega) = \langle S^{\alpha}_{k\omega} S^{\alpha}_{-k\omega} \rangle \quad \text{and} \quad D_0(k,\omega) = \langle \varepsilon_{k\omega} \varepsilon_{-k\omega} \rangle .$$

From the Fourier transformed versions of (5.28a) and (5.28b) we obtain

$$C_0(k,\omega) = \frac{2\Gamma_0 k^2}{\omega^2 + (\Gamma_0 k^2 (r_0 + k^2))^2} \tag{5.30a}$$

$$D_0(k,\omega) = \frac{2\gamma_0 k^2}{\omega^2 + (\gamma_0 C_0 k^2)^2} . \qquad (5.30b)$$

We notice that for a purely diffusive process the correlation functions have a Lorentzian shape centered about the origin with line widths proportional to k^2. In the long wavelength limit k = 0 the correlation functions are proportional to $\delta(\omega)$; again a consequence of the respective conservation laws. In Fig. 5.5 we have depicted the correlation function $C_0(k,\omega)$ as a function of ω .

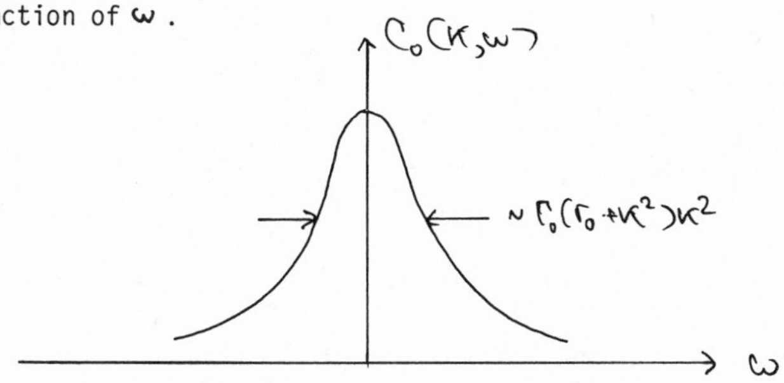

Fig. 5.5 The Lorentzian form of the spin correlation function $C_0(k,\omega)$

The non linear terms in the Langevin equations (5.27a) and (5.27b) give rise to interactions between the spin and energy fluctuations. By inspection of (5.27a) and (5.27b) and retaining only terms involving two fields we distinguish the mode couplings $\lambda_0 \sum_{\beta\gamma} \varepsilon^{\alpha\beta\gamma} S^\beta \nabla^2 S^\gamma$, $\Gamma_0 g_0 \nabla^2(\varepsilon S^\alpha)$, and $\eta_0 r_0 \nabla(\varepsilon \nabla S^\alpha)$ for the spin equation, and $\frac{1}{2}\gamma_0 g_0 \nabla^2 S^2$ for the energy equation. Here the only reversible contribution is the precessional term $\lambda_0 S \times \nabla^2 S$ arising from the microscopic spin dynamics; the remaining terms are all irreversible.

In order to evaluate the effects due to the mode coupling terms one now establishes a perturbation scheme based on iterating the inhomogeneous Langevin equations (5.27a) and (5.27b). Such a program was carried out in ref. 65. In the present context we disregard, for simplicity, the coupling to the energy fluctuations and consider in Fourier space the equation of motion for $S^\alpha_{k\omega}$,

$$(-i\omega + \Gamma_0 k^2(r_0+k^2)) S^\alpha_{k\omega} = f^\alpha_{k\omega} + \lambda_0 \sum_{\beta\gamma} \varepsilon^{\alpha\beta\gamma} \int p^2 S^\beta_{k-p\,\omega-\nu} S^\gamma_{p\nu} \frac{d^dp\,d\nu}{(2\pi)^{d+1}} + h^\alpha_{k\omega}$$

$$\langle f^\alpha_{k\omega} f^\beta_{p\nu} \rangle = 2\Gamma_0 \delta^{\alpha\beta} k^2 (2\pi)^{d+1} \delta(\omega+\nu)\, \delta^d(k+p) , \qquad (5.31)$$

where we have added a non random force $h^{\alpha}_{k\omega}$ in order to define the general response function $G(k,\omega)$ by $\delta\langle S^{\alpha}_{k\omega}\rangle = G(k,\omega)h^{\alpha}_{k\omega}$. It is convenient to represent the terms in the perturbation expansion by means of diagrams[36,75,76]. The lines in the diagrams have arrows indicating the direction of iteration. There are two types of lines, circled and uncircled. An uncircled line ⟶ corresponds to the unperturbed propagator $G_0(k,\omega) = 1/(-i\omega + \Gamma_0 k^2(r_0+k^2))$; a circled line , where the circle indicates a noise average $2\Gamma_0 k^2$, yields the unperturbed correlation function $C_0(k,\omega) = 2\Gamma_0 k^2/(\omega^2+(\Gamma_0 k^2(r_0+k^2))^2)$. A vertex in the diagram has one incoming line specified by frequency, wavenumber and spin index $(\omega+\nu, k+p, \alpha)$, and two outgoing lines (ω, k, β) and (ν, p, γ), and gives the contribution $-\lambda_0\,\varepsilon^{\alpha\beta\gamma}\,p^2$. Conserving total frequency and total wavenumber at each vertex, one sums over all spin indices and integrates over all frequencies and wavenumbers. The diagrammatic contributions fall in three classes: Self energy corrections, noise corrections, and vertex corrections. The self energy corrections $\Sigma(k,\omega)$ have one incoming and one outgoing line, the noise corrections $N(k,\omega)$ two incoming lines, and the vertex correction $\Lambda(k\omega, p\nu)$ one incoming and two outgoing lines.

By inspection of (5.31) it follows that the response function $G(k,\omega)$ has the general form (see also refs. 21 and 67)

$$G(k,\omega) = \frac{1}{-i\omega + \Gamma_0 k^2(r_0+k^2) - \Sigma(k,\omega)}, \qquad (5.32)$$

where $\Sigma(k,\omega)$ is the diagrammatically defined self energy. To second order in λ_0 the self energy is given by the diagram

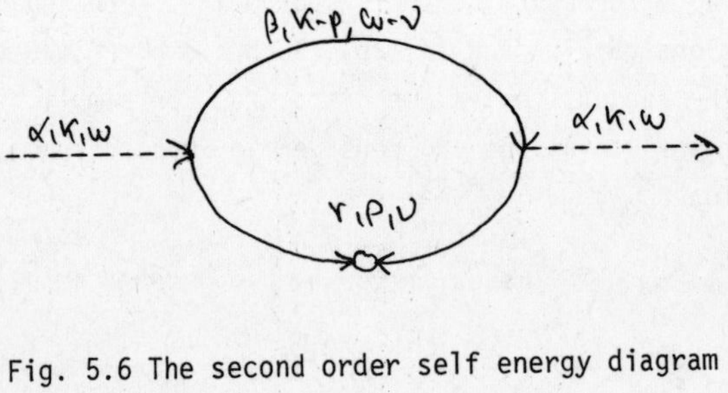

Fig. 5.6 The second order self energy diagram

Applying the diagrammatic rules we obtain the contribution

$$\Sigma(k,\omega) =$$

$$2\lambda_0^2 \int \frac{d^dp\,d\nu}{(2\pi)^{d+1}} \frac{[p^2-(k-p)^2][k^2-p^2]2\Gamma_0 p^2}{[-i(\omega-\nu)+\Gamma_0(k-p)^2(r_0+(k-p)^2)][\nu^2+(\Gamma_0 p^2(r_0+p^2))^2]} .$$

The propagator has a pole at $\omega+i\Gamma_0(k-p)^2(r_0+(k-p)^2)$ and the Lorentzian factor poles at $\pm i\Gamma_0 p^2(r_0+p^2)$. Closing the contour in the lower half plane, integration over ν yields

$$\Sigma(k,\omega) = 2\lambda_0^2 \int \frac{d^dp}{(2\pi)^d} \frac{[p^2-(k-p)^2][k^2-p^2]}{[-i\omega+\Gamma_0 p^2(r_0+p^2)+\Gamma_0(k-p)^2(r_0+(k-p)^2)][r_0+p^2]} .$$

Symmetrising $p \leftrightarrow k-p$ and introducing $p \leftrightarrow p+k/2$ we obtain to leading order in k

$$\Sigma(k,\omega) = -4\lambda_0^2 k^2(r_0+k^2)\frac{1}{d}\int \frac{d^dp}{(2\pi)^d} \frac{p^2}{[-i\omega+\Gamma_0(r_0+p^2)(2p^2+\frac{k^2}{2})][r_0+p^2]^2} ,$$

where the dimension d enters explicitly since $\int (k\cdot p)^2 f(p^2)d^dp = \sum_{ij=1}^{d} k^ik^j \int p^ip^jf(p^2)d^dp = \frac{k^2}{d}\int p^2 f(p)d^dp$. The area of a unit sphere in d dimensions is $2\pi^{d/2}/\Gamma(d/2)$. Notice that for d=3 we obtain, since $\Gamma(3/2) = \pi^{\frac{1}{2}}/2$, 4π. Introducing $K_d = (2\pi)^{-d}2\pi^{d/2}/\Gamma(d/2)$ the second order correction to the self energy is

$$\Sigma(k,\omega) = -4\lambda_0^2(r_0+k^2)k^2\frac{K_d}{d}\int_0^\Lambda dp \frac{p^{d+1}}{[r_0+p^2]^2[-i\omega+\Gamma_0(r_0+p^2)(2p^2+\frac{k^2}{2})]} ,$$

where we have introduced an ultraviolet wavenumber cut off Λ implicit in the "coarse grained" hydrodynamical description. By insertion of $\Sigma(k,\omega)$ in the expression (5.32) we notice that, in accordance with the general formalism discussed in Section 5.4 and 5.5, the reversible mode coupling does not affect the static properties, and we identify the correction to the damping coefficient Γ_0,

$$\Delta\Gamma(k,\omega) = 4\lambda_0^2 \frac{K_d}{d}\int_0^\Lambda dp \frac{p^{d+1}}{[r_0+p^2]^2[-i\omega+\Gamma_0(r_0+p^2)(2p^2+\frac{k^2}{2})]} . \qquad (5.33)$$

It follows from our discussion and from general principles[21] that the spin correlation function

$$C(k,\omega) = \langle S_{k\omega} S_{p\nu} \rangle / (2\pi)^{d+1} \delta(\omega+\nu)\, \delta^d(k+p)$$

has the general form $C(k,\omega) = 2\mathrm{Re}G(k,\omega)/r_0$, i.e.,

$$C(k,\omega) = \frac{2}{r_0} \mathrm{Re}\left(\frac{1}{-i\omega + \Gamma(k,\omega) r_0 k^2} \right) \tag{5.34}$$

where $\Gamma(k,\omega) = \Gamma_0 + \Delta\Gamma(k,\omega)$ is the frequency and wavenumber dependent damping coefficient. The relationship (5.34) expresses the well-known fluctuation-dissipation theorem[20,21]. A correction to the damping coefficient (dissipation) must therefore be balanced by a corresponding correction $N(k,\omega)$ to the noise average $\Gamma_0 k^2$ (fluctuation). To second order in λ_0 we have indeed the noise correction $N(k,\omega)$,

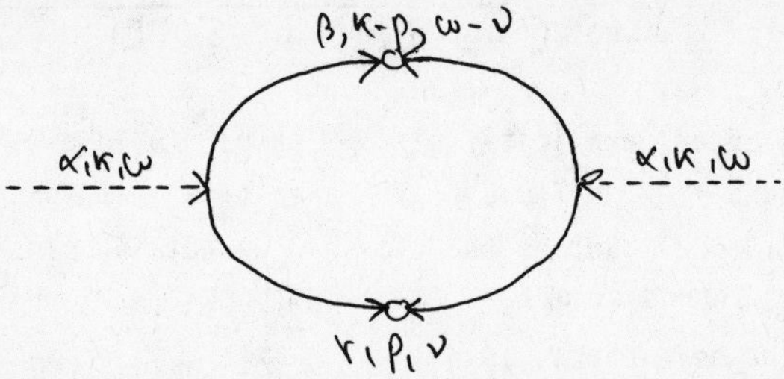

Fig. 5.7 The second order noise diagram

which according to the diagrammatic rules yields the contribution,

$$N(k,\omega) = \lambda_0^2 \int \frac{d^d p\, d\nu}{(2\pi)^{d+1}} \frac{((k-p)^2 - p^2)^2\, 4\Gamma_0^2 (k-p)^2 p^2}{(\nu^2 + (\Gamma_0 p^2 (r_0 + p^2))^2)\,((\omega-\nu)^2 + (\Gamma_0 (k-p)^2 (r_0 + (k-p)^2))^2)} .$$

Carrying out the integration over ν and symmetrising we can express $N(k,\omega)$ in the form $2k^2 \mathrm{Re}(\Delta\Gamma(k,\omega))$, consistent with the fluctuation-dissipation theorem (5.34) to second order in λ_0.

Iterating (5.31) we obtain to third order in λ_0

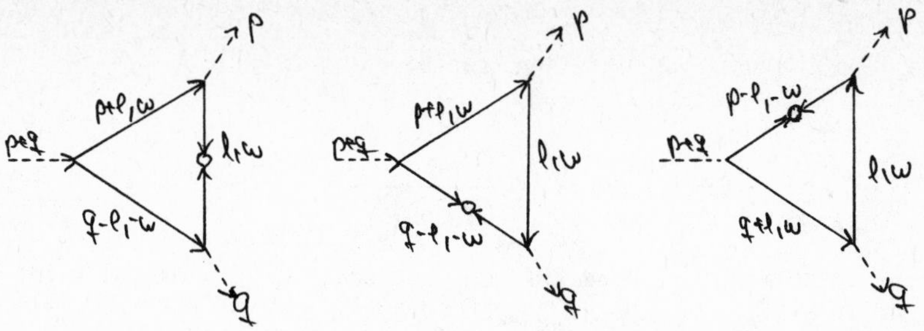

Fig. 5.8 The third order vertex diagrams

In general the vertex correction $\Lambda(p\nu, k\omega)$ is frequency and wavenumber dependent. Here we consider, however, only the static long wavelength limit $k = 0$ and $\omega = 0$. Applying the diagrammatic rules we have

$$\Lambda(p0, q0) = \lambda_0^3 \int \frac{d^dk}{(2\pi)^d} \frac{d\omega}{(2\pi)} \times$$

$$\Bigg(\frac{((q-k)^2-(p+k)^2)(k^2-p^2)\, 2\Gamma_0 k^2}{(-i\omega+\Gamma_0(p+k)^2(r_0+(p+k)^2))(i\omega+\Gamma_0(q-k)^2(r_0+(q-k)^2))(\omega^2+(\Gamma_0 k^2(r_0+k^2))^2)}$$

$$+ \frac{((q-k)^2-(p+k)^2)(k^2-p^2)((q-k)^2-q^2)\, 2\Gamma_0 (q-k)^2}{(-i\omega+\Gamma_0(p+k)^2(r_0+(p+k)^2))(-i\omega+\Gamma_0 k^2(r_0+k^2))(\omega^2+(\Gamma_0(q-k)^2(r_0+(q-k)^2))^2)}$$

$$+ \frac{((q+k)^2-(p-k)^2)(q^2-k^2)(p^2-(p-k)^2)\, 2\Gamma_0 (p-k)^2}{(-i\omega+\Gamma_0(q+k)^2(r_0+(q+k)^2))(-i\omega+\Gamma_0 k^2(r_0+k^2))(\omega^2+(\Gamma_0(p+k)^2(r_0+(p+k)^2))^2)} \Bigg)$$

Performing the frequency integration and after a fair amount of algebraic manipulations the above contributions can be expressed in the form $\Delta\lambda(p^2-q^2)$, where the static vertex correction $\Delta\lambda$ is given by

$$\Delta\lambda = -\frac{\lambda_o^3 r_o}{\Gamma_o^2}\,\frac{K_d}{d}\int_0^{\Lambda} dp\,\frac{p^{d-1}}{(r_o+p^2)^3}\,. \tag{5.35}$$

We notice that the vertex correction $\Delta\lambda$ vanishes at the critical point $r_0 = 0$ for $d > 6$ (see ref.75).

5.7 Correction to Hydrodynamics - Long Time Tails

In the previous section we evaluated the second order correction to the wavenumber and frequency dependent transport coefficient $\Gamma(k,\omega)$ for the Heisenberg paramagnet in the long wavelength limit, taking into account the reversible mode coupling $\lambda_0 \nabla(S\times\nabla S))$. In ref. 65 it was shown that the non linear irreversible drift term $\eta_0 r_0 \nabla(\varepsilon\nabla S)$ yields a similar correction (of opposite sign) to the transport coefficient,

$$\Delta\Gamma(k,\omega) = -\frac{\eta_o^2\Gamma_o}{c_o}\,\frac{K_d}{d}\int_0^{\Lambda}\frac{p^{d+1}\,dp}{(\Gamma_o r_o+\gamma_o c_o)p^2+(\Gamma_o r_o\gamma_o c_o/(\Gamma_o r_o+\gamma_o c_o))k^2-i\omega} \tag{5.36}$$

Since the coupling to the energy fluctuations does not give rise to qualitative changes, we confine in the following our discussion to the pure spin case.

For small k and ω the correction $\Delta\Gamma(k,\omega)$ given by (5.33) is controlled by the branch point $\omega = -i\Gamma_0 r_0 k^2/2$. In the long time limit one finds

$$\Delta\Gamma(k,t) \simeq \left(\frac{\lambda_o}{r_o}\right)^2\left(\frac{K_d}{d}\right)|t|^{-(1+\frac{d}{2})}\exp\left(-\Gamma_o r_o k^2\frac{|t|}{2}\right) \tag{5.37}$$

i.e., the correction $\Delta\Gamma$ decays exponentially with a life time $\sim 1/\Gamma_0 r_0 k^2$. At zero wavenumber $k = 0$ the branch point is at the origin and $\Delta\Gamma(k=0,t)$ falls off algebraically as $|t|^{-(1+d/2)}$, in contrast to the assumptions underlying linearised hydrodynamics, according to which $\Delta\Gamma(k=0,t)$ should decay rapidly in a microscopic time. In frequency space $\Delta\Gamma(k=0,\omega)$ varies as $\omega^{d/2}$ for $d > 0$ and as $\log\omega$ for $d = 0$. In three dimensions the time dependent transport coefficient exhibits a $t^{-5/2}$ power law behaviour.

Long time tails indicative of non analytic corrections to hydrodynamics were first studied for the case of transverse velocity correlations in a fluid (see, for instance, the recent review ref. 77). Here a variety of calculations indicate that the correction to the ω and k dependent transport coefficient, i.e., the kinematical viscosity $\nu(k,\omega)$, for k = 0 behaves as $\omega^{(d-2)/2}$ for $d > 2$ and as $\log\omega$ for d = 2. The correction to hydrodynamics thus diverges in two dimensions. The associated long time tail behaves as $t^{-d/2}$ producing the well-known $t^{-3/2}$ fall off in three dimensions for the auto velocity correlation function. The difference in power law and crossover dimension are associated with the form of Navier-Stokes equation for an incompressible fluid[71]

$$\frac{d\bar{v}}{dt} = -(\bar{v}\nabla)\bar{v} + \nu_0\nabla^2\bar{v}$$

compared to the spin equation

$$\frac{d\bar{S}}{dt} = \lambda_0 \bar{S}\times\nabla^2\bar{S} + \Gamma_0 r_0 \nabla^2\bar{S} .$$

Since the mode coupling in the fluid case $(\bar{v}\nabla)\bar{v}$ has only one gradient in contrast to $\bar{S}\times\nabla^2\bar{S}$ in the spin case the crossover dimension changes from d = 0 to d = 2.

The prospect of observing the non analytic corrections to spin hydrodynamics and the associated long time tails is, unfortunately, somewhat remote. In a typical neutron scattering experiment one measures primarily the spin correlation function[21]. From the form of the spectral representation (5.3)

$$C(k,\omega) = \langle S_{k\omega} S_{-k-\omega}\rangle = \frac{2k^2 \,\mathrm{Re}\,\Gamma(k,\omega)}{(\omega - r_0 k^2 \,\mathrm{Im}\,\Gamma(k,\omega))^2 + (r_0 k^2 \,\mathrm{Re}\,\Gamma(k,\omega))^2} \quad (5.38)$$

it follows that $C(k,\omega)$ has a singular behaviour of the same type as $\Gamma(k,\omega)$ for $k,\omega \to 0$. However, any measurement must be made at finite k since $C(k=0,\omega) \propto \delta(\omega)$ because of the conservation law, and the power law behaviour is masked by the exponential factor in (5.37).

At k = 0 the spectral form (5.38) implies that

$$\mathrm{Re}\,\Gamma(k=0,\omega) = \frac{1}{2d}\langle J^\alpha_{0\omega} J^\alpha_{0-\omega}\rangle ,$$

where the spin current J^α is related to S^α by the continuity equation

$\omega S^{\alpha}_{k\omega} = kJ^{\alpha}_{k\omega}$. The current correlation function at k = 0 should therefore fall off like $t^{-(1+d/2)}$ and in principle this effect could be seen in a computer simulation.

We finally remark that the hydrodynamical corrections due to the precessional term $S \times \nabla^2 S$ has also been considered by Månson[78]. A treatment including the coupling to energy fluctuations has independently been carried out by Borckmans et al.[79] using the mode coupling formalism of Kawasaki.

5.8 Renormalisation Group Treatment

In recent work Forster et al.[80.81] have applied modern renormalisation group theory to a study of the hydrodynamics of an incompressible fluid. Here we apply these methods to the paramagnet. For simplicity we disregard the energy fluctuations and consider the non-linear equation

$$\frac{d\bar{S}}{dt} = \lambda_0 \bar{S} \times \nabla^2 \bar{S} + \Gamma_0 \nabla^2 (r_0 - \nabla^2) \bar{S} + \bar{f}$$

$$\langle f^{\alpha}(x,t) f^{\beta}(x',t') \rangle = -2\Gamma_0 \delta^{\alpha\beta} \delta(t-t') \nabla^2 \delta^d(x-x') . \quad (5.39)$$

In the previous section we calculated to second order in λ_0 the corrections $\Delta\Gamma(k,\omega)$ and $\Delta\lambda$ to the damping coefficient and the mode coupling. In the static long wavelength limit we have from (5.33) and (5.35)

$$\Delta\Gamma = \frac{2\lambda_0^2}{\Gamma_0} \frac{K_d}{d} \int_0^{\Lambda} dp \frac{p^{d-1}}{(r_0 + p^2)^3} ,$$

$$\Delta\lambda = -\frac{\lambda_0^3 r_0}{\Gamma_0^2} \frac{K_d}{d} \int_0^{\Lambda} dp \frac{p^{d-1}}{(r_0 + p^2)^3} .$$

For the d = 0 the corrections are logarithmically divergent, for $d < 0$ algebraically divergent. The divergence arises from the lower integration limit, i.e., the far infrared limit. In order to investigate the singular behaviour in the long wavelength limit we apply a renormalisation group analysis, following

The correction $\Delta\Gamma$ arises from the self energy diagram in Fig.5.6. The integration from 0 to the cut off Λ implies that we are including all intermediate processes generated by the mode coupling $\lambda_0\bar{S}\times\nabla^2\bar{S}$. The essence of the renormalisation group approach is now to average over the short wavelength degrees of freedom in steps and thereby disentangle the long wavelength dynamics of the problem. Following ref. 81 we integrate over degrees of freedom in a wavenumber shell $\Lambda\exp(-\ell) < k < \Lambda$, where ℓ is a logarithmic scale factor. This procedure is non linear and non perturbative and will in general produce higher order mode coupling terms. Discarding those terms and retaining only the leading contributions the partially integrated equation of motion takes the form

$$\frac{d\bar{S}^<}{dt} = (\lambda_0+\Delta\lambda)\bar{S}^<\times\nabla^2\bar{S}^< + (\Gamma_0+\Delta\Gamma)\nabla^2(r_0-\nabla^2)\bar{S}^< + \bar{f}^<$$

$$\langle f^{\alpha<}(x,t)\, f^{\beta<}(x',t')\rangle = -2(\Gamma_0+\Delta\Gamma)\delta^{\alpha\beta}\delta(t-t')\nabla^2\delta^d(x-x') .$$

We have here to second order in λ_0 averaged over the Fourier components of the noise $f_k^\alpha(t)$ in the wavenumber shell $\Lambda\exp(-\ell) < k < \Lambda$. The field $\bar{S}^<$ and noise $\bar{f}^<$, consequently, only have Fourier components in the interval $0 < k < \Lambda\exp(-\ell)$. The corrections $\Delta\lambda$ and $\Delta\Gamma$ are accordingly given by

$$\Delta\Gamma = \frac{2\lambda_0^2}{\Gamma_0}\frac{K_d}{d}\int_{\Lambda\exp(-\ell)}^{\Lambda} dp\, \frac{p^{d-1}}{(\Gamma_0+p^2)^3}$$

$$\Delta\lambda = -\frac{\lambda_0^3\Gamma_0}{\Gamma_0^2}\frac{K_d}{d}\int_{\Lambda\exp(-\ell)}^{\Lambda} dp\, \frac{p^{d-1}}{(\Gamma_0+p^2)^3}$$

or, choosing the cut off $\Lambda \ll r_0^{1/2}$,

$$\Delta\Gamma = \frac{\lambda_0^2}{\Gamma_0}\left[\frac{1-\exp(-d\ell)}{d}\right]A_d \quad \text{and} \quad \Delta\lambda = -\frac{\lambda_0^3}{\Gamma_0^2}\left[\frac{1-\exp(-d\ell)}{d}\right]B_d ,$$

where $A_d = 2K_d\Lambda^d/dr_0^3$ and $B_d = K_d\Lambda^d/dr_0^2$ are positive constants, $K_d = (2\pi)^{-d}2\pi^{d/2}/\Gamma(d/2)$. In Fig. 5.9 we have shown the wavenumber shell integration

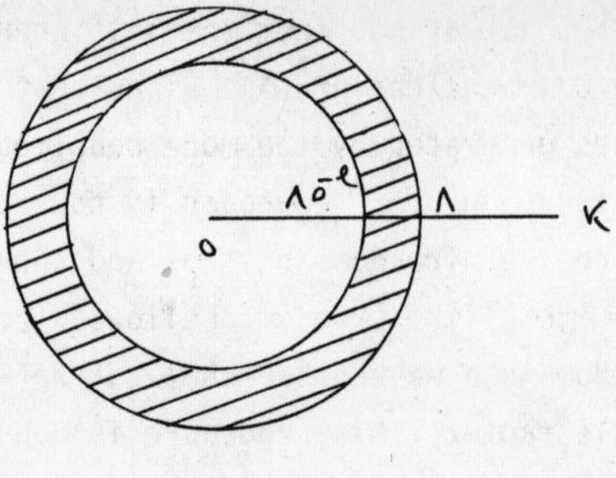

Fig. 5.9 The wavenumber shell integration. The shaded area shows the averaged degrees of freedom

The next step in the renormalisations group approach consists in rescaling the wavenumber k, the field $\bar{S}^{\alpha}_{k\omega}$, and the frequency ω such that 1) the scaled wavenumber k' runs over the original interval $0 < k' < \Lambda$ and 2) the scaled equation of motion takes the same form as the original one. Introducing $k' = k\exp(\ell)$, $\bar{S}^{<}_{k\omega} = \xi(\ell)\bar{S}'_{k'\omega'}$, $\bar{f}_{k\omega} = \xi(\ell)\exp(-\alpha(\ell))\bar{f}'_{k'\omega'}$, and $\omega' = \omega\exp(\alpha(\ell))$ we obtain, by insertion in the partially integrated equation of motion, for $\bar{S}^{<}$ in Fourier space

$$-i\omega'\bar{S}'_{k'\omega'} = [(\lambda_0+\Delta\lambda)\xi(\ell)\exp(-(d+2)\ell)](\bar{S}'\times\nabla'^2\bar{S}')_{k'\omega'} +$$

$$[(\Gamma_0+\Delta\Gamma)\exp(\alpha(\ell))\exp(-2\ell)]k'^2[r_0+\exp(-2\ell)k'^2]\bar{S}_{k'\omega'} + \bar{f}'_{k'\omega'}$$

$$\langle f'_{k'\omega'} f'_{p'\nu'}\rangle = [\xi(\ell)^{-2}\exp(2\alpha(\ell))\exp(d\ell)][2(\Gamma_0+\Delta\Gamma)\exp(\alpha(\ell))\exp(-2\ell)]\times$$

$$\delta^{\alpha\beta}(2\pi)\delta(\omega'+\nu')(2\pi)^d\delta^d(k'+p')k'^2$$

In order to preserve the Einstein relation, i.e., the fluctuation-dissipation theorem, under scaling we must have $\xi(\ell)^{-2}\exp(2\alpha(\ell))\exp(d\ell)=1$, i.e., we choose the field scaling $\xi(\ell)=\exp(\alpha(\ell)+\ell d/2)$. The new equation of motion describing the same long wavelength-low frequency dynamics thus takes the form

$$\frac{d\vec{S}'}{dt'} = \lambda'\vec{S}'\times\nabla'^2\vec{S} + \Gamma'\nabla'^2(r_0 - \exp(-\ell)\nabla^2)\vec{S}' \tag{5.40}$$

$$\langle f^{\alpha'}(x',t')f^{\beta'}(y,\tau')\rangle = -2\Gamma'\delta^{\alpha\beta}\delta(t'-\tau')\nabla'^2\delta^d(x'-y'),$$

where $\lambda' = (\lambda_0+\Delta\lambda)\exp(\alpha(\ell)-d\ell/2-2\ell)$ and $\Gamma' = (\Gamma_0+\Delta\Gamma)\exp(\alpha(\ell)-2\ell)$ are the scaled mode coupling and damping. Inserting the corrections $\Delta\Gamma$ and $\Delta\lambda$ we obtain the renormalisation group equations

$$\Gamma' = \left(\Gamma_0 + A_d\frac{\lambda_0^2}{\Gamma_0}\frac{1-\exp(-d\ell)}{d}\right)\exp(\alpha(\ell)-2\ell) \tag{5.41a}$$

$$\lambda' = \left(\lambda_0 - B_d\frac{\lambda_0^3}{\Gamma_0^2}\frac{1-\exp(-d\ell)}{d}\right)\exp\left(\alpha(\ell)-\frac{d\ell}{2}-2\ell\right) \tag{5.41b}$$

We stress that the mode coupling equation (5.40) describes the same low frequency-long wavelength hydrodynamics as (5.39) since we have only averaged over the short wavelength modes in the interval $\Lambda\exp(-\ell) < k < \Lambda$. To second order in λ_0 the result of this procedure is to change the coupling λ_0 and the damping Γ_0 according to (5.41a) and (5.41b).

The final step in the renormalisation group approach amounts to iterating the above procdure, i.e., we successively integrate out the short wave length modes, and search for fixed points of the renormalisation group equations (5.41a) and (5.41b). This program is most easily carried out by considering an infinitesimal scaling and converting (5.41a) and (5.41b) to differential equations. The case $\ell = 0$ corresponds to no scaling at all, i.e., $\alpha(\ell) = 0$. For small ℓ we write $\alpha(\ell) = z(\ell)\ell$, i.e., $\alpha(\ell) = \int_0^\ell z(\ell')d\ell'$. Expanding the exponentials in (5.41a) and (5.41b) for small ℓ and setting $\Gamma' = \Gamma(\ell)+d\Gamma(\ell)$ and $\lambda' = \lambda(\ell)+d\lambda(\ell)$ we obtain renormalisation group equations in differential form,

$$\frac{d\Gamma(\ell)}{d\ell} = \Gamma(\ell)\left(-2+z(\ell)+A_d\left(\frac{\lambda(\ell)}{\Gamma(\ell)}\right)^2\right) \tag{5.42a}$$

$$\frac{d\lambda(\ell)}{d\ell} = \lambda(\ell)\left(-2+z(\ell)-\frac{d}{2}-B_d\left(\frac{\lambda(\ell)}{\Gamma(\ell)}\right)^2\right) \tag{5.42b}$$

with the inital values $\lambda(0) = \lambda_0$ and $\Gamma(0) = \Gamma_0$.

The non linear equations (5.42a) and (5.42b) describe a set of trajectories, a so-called flow[35], in two dimensional parameter space $(\Gamma(\ell), \lambda(\ell))$. The equations are, however, reduced by noting that we can choose the frequency scaling $z(\ell)$ such that $\Gamma(\ell) = \Gamma_0$ for all ℓ, i.e., $z(\ell) = 2 - A_d(\lambda(\ell)/\Gamma_0)^2$. Inserting $z(\ell)$ in (5.42b) we obtain a single equation for the reduced mode coupling $\bar{\lambda}(\ell) = \lambda(\ell)/\Gamma_0$,

$$\frac{d\bar{\lambda}(\ell)}{d\ell} = -\frac{d}{2}\bar{\lambda}(\ell) - (A_d + B_d)\bar{\lambda}(\ell)^3 \tag{5.43}$$

with the solution

$$\bar{\lambda}(\ell) = \lambda_0 \frac{\exp(-\frac{d\ell}{2})}{\left(1 + 2\frac{A_d+B_d}{d}\bar{\lambda}_0^2(1-\exp(-d\ell))\right)^{\frac{1}{2}}} . \tag{5.44}$$

The fixed points $\bar{\lambda}^*$ of the renormalisation group equation (5.43) are found by setting $d\bar{\lambda}(\ell)/d\ell = 0$, i.e., $\frac{d}{2}\bar{\lambda}^* + (A_d+B_d)\bar{\lambda}^{*3} = 0$. In physical dimensions $d > 0$ we have only one fixed point at $\bar{\lambda}^* = 0$; for $d < 0$ we obtain an additional fixed point $\bar{\lambda}^* = (|d|/2(A_d+B_d))^{1/2}$, or to order $|d|^{3/2}$ $\bar{\lambda}^* = (|d|/2(A_0+B_0))^{1/2}$, where $A_0 = 2/r_0^3$ and $B_0 = 1/r_0^2$. By inspection of (5.44) we notice that for $d > 0$ the fixed point $\bar{\lambda}^*$ is stable, whereas for $d < 0$ the stable fixed point $\bar{\lambda}^* = (|d|/2(A_0+B_0))^{1/2}$ emerges and $\bar{\lambda}^* = 0$ becomes unstable. In Fig. 5.1o we have shown the fixed points as a function of the dimension d.

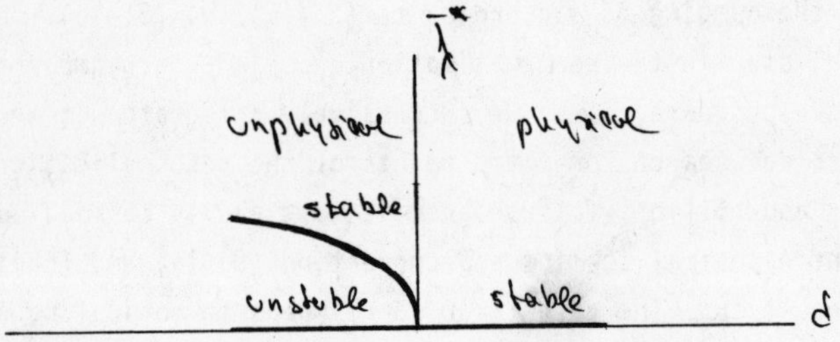

Fig. 5.10 The fixed points as a function of the dimension d

In Fig. 5.11 and Fig. 5.12 we have shown the trajectories in the two cases $d > 0$ and $d < 0$.

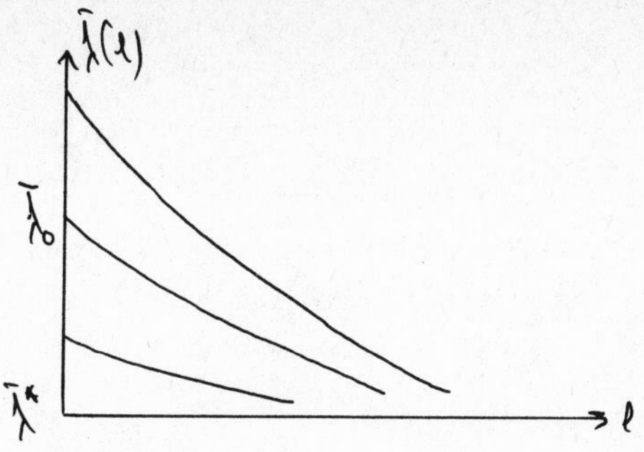

Fig. 5.11 The trajectories of $\bar{\lambda}(\ell)$ for different inital values $\bar{\lambda}_0$ for $d > 0$

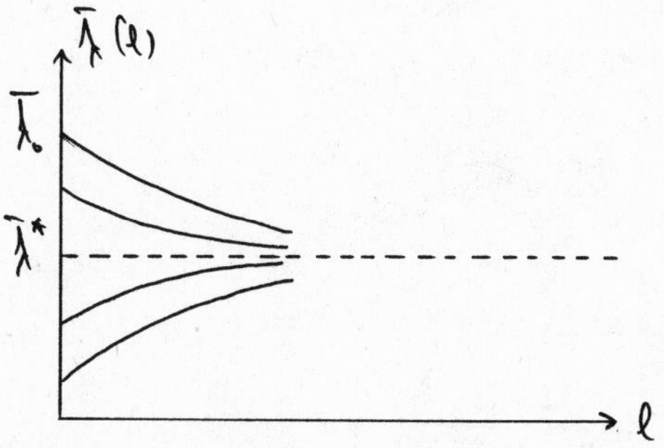

Fig. 5.12 The trajectoris of $\bar{\lambda}(\ell)$ for different initial values $\bar{\lambda}_0$ for $d < 0$

The preceeding analysis also enables us to establish the scaling form of the spin correlation function $C(k,\omega)$. We have

$$C(k,\omega) = \langle S^\alpha_{k\omega} S^\alpha_{p\nu} \rangle / (2\pi)^{d+1} \delta(\omega+\nu)\, \delta^d(k+p) .$$

For $k < \Lambda \exp(-\ell)$ we can, of course, evaluate $C(k,\omega)$ either from the original equation of motion (5.39) or from the renormalised one (5.40), describing the same physics in the long wavelength-low frequency limit. Following ref. 81 which gives a particularly lucid discussion we obtain, using $S' = \xi(\ell)^{-1} S^<$, $k' = k\exp(\ell)$, $\omega' = \omega\exp(\alpha(\ell))$, and $\xi(\ell) = \exp(\alpha(\ell) + \ell d/2)$,

$$C(k,\omega;\bar{\lambda}_0) = \langle S^<_{k\omega} S^<_{p\nu}\rangle/(2\pi)^{d+1}\delta(\omega+\nu)\,\delta^d(k+p)$$

$$= \xi(\ell)^2\langle S'_{k'\omega'} S'_{p'\nu'}\rangle/(2\pi)^{d+1}\delta(\omega+\nu)\,\delta^d(k+p)$$

$$= \exp(\alpha(\ell))\, C(k',\omega';\bar{\lambda}(\ell))$$

i.e., we have the scaling relationship

$$C(k,\omega,\bar{\lambda}_0) = \exp\left(\int_0^\ell z(\ell')d\ell'\right) C\left(k\exp(\ell),\omega\exp\left(\int_0^\ell z(\ell')d\ell'\right);\bar{\lambda}(\ell)\right) \quad (5.45)$$

For d >0 and ℓ large $\lambda(\ell)$ approaches the fixed point $\lambda^* = 0$. Since $z(\ell) = 2 - A_d(\lambda(\ell)/\Gamma_0)^2$, $z(\ell)$ approaches 2, i.e., $\exp(\int_0^\ell z(\ell')d\ell') \to \exp(2\ell)$. Consequently, for large ℓ and for k less than $\Lambda\exp(-\ell)$ the relationship (5.45) implies that

$$C(k,\omega) = \Phi(\omega/k^2)/k^2,$$

where $\Phi(x)$ is a scaling function. This is the prediction of conventional linearised spin hydrodynamics provided we take

$$\Phi(x) = \frac{2\Gamma}{x^2 + (\Gamma r_0)^2} \quad (5.46)$$

in accordance with the form of (5.30a). Here Γ is the renormalised damping coefficient.

In order to evaluate Γ within the renormalisation group approach we notice that for large ℓ we have $z(\ell) = 2 - A_d\,\bar{\lambda}(\ell)^2$ or

$$\int_0^\ell z(\ell')d\ell' = 2\ell - A_d\int_0^\ell \bar{\lambda}(\ell')^2 d\ell'$$

Denoting the correction $A_d\int_0^\infty \bar{\lambda}(\ell')d\ell'$ by Δ we obtain with exponentially small error from the scaling relationship (5.45)

$$C(k,\omega;\bar{\lambda}_0) = \exp(2\ell-\Delta)\,C(k\exp(\ell),\omega\exp(2\ell-\Delta);\bar{\lambda}(\ell)).$$

Assuming $\bar{\lambda}_0$ small and $\bar{\lambda}(\ell) \simeq 0$, and inserting the scaling form (5.46) of linearised hydrodynamics we have

$$\frac{2\Gamma k^2}{\omega^2 + (\Gamma \Gamma_0 k^2)^2} = \frac{2\Gamma_0 k^2 \exp(2\ell)}{\omega^2 \exp(4\ell - 2\Delta) + (\Gamma_0 \Gamma_0 k^2)^2 \exp(4\ell)}$$

and we infer $\Gamma/\Gamma_0 = \exp(\Delta)$. The correction $\Delta = A_d \int_0^\infty \bar{\lambda}(\ell')^2 d\ell'$ is now evaluated by means of (5.44). To leading order in $\bar{\lambda}_0$, $\Delta = A_d \bar{\lambda}_0^2/d$, and we obtain $\Gamma = \Gamma_0 + A_d \lambda_0^2/d\Gamma_0$ in accordance with the previous calculations in Section 5.6.

From the general spectral form (5.34) and Eq.(5.45) it is easy to derive a scaling relation for the frequency and wave-number dependent damping coefficient $\Gamma(k,\omega)$. By insertion we obtain

$$\Gamma(k,\omega;\bar{\lambda}_0) = \exp(2\ell - \alpha(\ell))\, \Gamma(k\exp(\ell), \omega\exp(\alpha(\ell)); \bar{\lambda}(\ell)) \tag{5.47}$$

The singular corrections to the damping coefficient $\Gamma(k,\omega)$ evaluated in Section 5.7 can now be determined within the present renormalisation group analysis by examining the approach to the fixed point $\bar{\lambda}^* = 0$ for $d > 0$.

In the long wavelength limit $k = 0$ we have from 5.47)

$$\Gamma(0,\omega;\bar{\lambda}_0) = \exp(2\ell - \alpha(\ell))\, \Gamma(0, \omega\exp(\alpha(\ell)); \bar{\lambda}(\ell))$$

For ℓ large $\alpha(\ell) \simeq 2\ell$ and $\bar{\lambda}(\ell) \simeq \bar{\lambda}_0 \exp(-d\ell/2)$, i.e., $\Gamma(0,\omega,\bar{\lambda}_0) = \Gamma(0, \omega\exp(2\ell), \bar{\lambda}_0 \exp(-d\ell/2))$. Choosing $\ell = \ell^*$ such that $\omega\exp(2\ell^*)$ is of order unity, i.e., $\exp(\ell^*) \simeq \omega^{-1/2}$, we obtain $\Gamma(0,\omega,\bar{\lambda}_0) = \Gamma(0,1,\bar{\lambda}_0 \omega^{d/4})$. Furthermore, choosing ω small enough the effective coupling $\bar{\lambda}(\ell^*) = \bar{\lambda}_0 \omega^{d/4}$ becomes as small as desired and we can do perturbation theory. Since only even powers in $\bar{\lambda}$ enters we have $\Gamma(0,\omega,\bar{\lambda}_0) = \Gamma_0(1 + \text{const.}\,\bar{\lambda}(\ell^*)^2)$ and we infer the singular correction $\Gamma(0,\omega,\bar{\lambda}_0)$,

$$\Gamma(0,\omega;\bar{\lambda}_0) = \Gamma_0((1 + \text{const.}\,\omega^{\frac{d}{2}}), \tag{5.48}$$

which gives rise to the long time tail

$$\Gamma(0,t;\bar{\lambda}_0) \simeq \text{const.}\, t^{-1-\frac{d}{2}} \tag{5.49}$$

In d = 3 we obtain in particular

$$\Gamma(0,t;\bar{\lambda}_0) \simeq \text{const.}\ t^{-\frac{5}{2}}$$

Similarly, for $\omega = 0$ we have for large ℓ $\Gamma(k,0,\bar{\lambda}_0) = \Gamma(k\exp(\ell),0,\bar{\lambda}_0\exp(-d\ell/2))$. Choosing $\ell=\ell^*$ such that $k\exp(\ell^*) = 1$, $\Gamma(k,0,\bar{\lambda}_0) = \Gamma(1,0,\bar{\lambda}_0 k^{d/2})$. For k sufficiently small the effective coupling $\bar{\lambda}(\ell^*) = \bar{\lambda}_0 k^{d/2}$ is small and we obtain the singular correction

$$\Gamma(k,0;\bar{\lambda}_0) \simeq \Gamma_0(1+\text{const.}\ k^d) \tag{5.50}$$

which in space yields the power law behaviour

$$\Gamma(x,0;\bar{\lambda}_0) \simeq \text{const.}\ x^{-2d}\ .$$

In d = 3 we have in particular

$$\Gamma(x,0;\bar{\lambda}_0) \simeq \text{const.}\ x^{-6}\ . \tag{5.51}$$

5.9 Concluding Remarks

The preceeding discussion of the hydrodynamics of the paramagnet is, neglecting energy fluctuations, based on the non linear Langevin equation[75]

$$\frac{dS}{dt} = -\lambda_0 S \times \frac{dF}{dS} + \Gamma_0 \nabla^2 \frac{dF}{dS} + \xi\ , \tag{5.52}$$

where the free energy functional F has the standard Ginzburg-Landau form[36]

$$F = \frac{1}{2}\int d^dx\left[r_0 S^2 + (\nabla S)^2 + \frac{1}{2}u_0 (S^2)^2 + \cdots\right] \tag{5.53}$$

obtained by coarse graining a microscopic classical or quantum mechanical Heisenberg model, replacing the length conditions $S^2 = 1$ by the non linear terms $\frac{1}{2}u_0(S^2)^2 + ..$ To second order in λ_0 we found singular corrections to the damping coefficient Γ_0 behaving as ω^d for $k = 0$ and k^d for $\omega = 0$. Since the corrections vanish in the long wavelength-low frequency limit $k,\omega \to 0$ we concluded that spin hydrodynamics holds for $d > 0$, i.e., in all physical dimensions.

In the light of recent work[82-85] it is, however, clear that the description of at least the static critical properties of the isotropic Heisenberg magnet in terms of the Ginzburg-Landau form (5.53) breaks down in $d \leqslant 2$ and has to be replaced by a theory based on the free energy functional

$$F = \frac{1}{2}K\int d^dx\,(\nabla S)^2, \quad S^2 = 1, \tag{5.54}$$

where the length condition enters as a constraint on the macroscopic level. This "fixed length" hydrodynamical description is implicit in ref. 82-85 and has been shown to be stable against small perturbations[86].

The dynamical description of a low dimensional isotropic paramagnet whose static properties are governed by the "fixed length" free energy (5.54) is, however, much more difficult. Although the reversible mode coupling term $\lambda_0 S \wedge \nabla^2 S$ preserves the length condition $S^2 = 1$, the damping term $\Gamma_0 r_0 \nabla^2 S$ and with it presumably the Langevin description must be abandoned. A recent attempt[87] has, however, been made for the dynamics of the XY model in $d \leqslant 2$, where the length condition $S_x^2 + S_y^2 = 1$ presents a weaker constraint and one still can set up a Langevin equation.

CHAPTER VI

SOLITONS AND MAGNONS IN THE CLASSICAL HEISENBERG CHAIN

In this final chapter we consider at zero temperature the dynamical properties of the classical one dimensional isotropic Heisenberg magnet in the long wavelength limit. This work, which is reported in ref.88, was prompted by the concluding remarks in Chapter V concerning the difficulty of setting up a hydrodynamical description for a low dimensional paramagnet. It became our point of view that in order to understand the statistical mechanics of the classical Heisenberg chain and eventually construct a proper dynamical description of the low temperature-low frequency-long wavelength properties, one must first understand the mechanics of the model, that is the fundamental dynamical modes.

It turns out that the classical Heisenberg chain, at least in the long wavelength limit, belongs to the class of completely integrable one dimensional Hamiltonian systems showing soliton behaviour and possessing an infinite set of constants of motion (see, for instance, refs. 89 and 90). Here we undertake a systematic investigation of the dynamics of the Heisenberg chain and proceed as far as displaying the soliton and magnon spectra and the series of constants of motion.

6.6 The Model

The dynamics of the classical one dimensional isotropic Heisenberg model in an applied constant magnetic field H_0 is characterised by the Hamiltonian

$$H = -J\sum_n \bar{S}_n \bar{S}_{n+1} - H_0 \sum_n \bar{S}_n ,$$

where $\bar{S}_n$ is a dimensionless classical spin scaled to unit length, $\bar{S}_n^2 = 1$. The positive nearest neighbour exchange coupling J and the magnetic field $\bar{H}_0$ are then of dimension energy. The site index n, n=1,2,...N, is associated with a one-dimensional lattice with lattice parameter a. Since $J > 0$, the spin configuration with lowest energy is ferromagnetic with all spins aligned in the direction of the field $\bar{H}_0$.

The Hamiltonian H being the generator of time translations[68] yields the equation of motion

$$\frac{d\bar{S}_n}{dt} = \{H, \bar{S}_n\},$$

where $\{A,B\}$ is the Poisson bracket[68]

$$\{A,B\} = \sum_{n,\alpha}\left(\frac{dA}{dp_n^\alpha}\frac{dB}{dq_n^\alpha} - \frac{dB}{dp_n^\alpha}\frac{dA}{dq_n^\alpha}\right),$$

defined in terms of an underlying canonical basis (q_n^α, p_n^α), $\alpha = x,y,z$. Since the spin $\bar{S}_n$ has the dynamical structure of an angular momentum, $S_n^\alpha = \sum_{\beta\gamma}\varepsilon^{\alpha\beta\gamma} q_n^\beta p_n^\gamma$, we immediately infer the non canonical Poisson bracket relations[68],

$$\{S_n^\alpha, S_m^\beta\} = -\delta_{nm}\sum_\gamma \varepsilon^{\alpha\beta\gamma} S_n^\gamma .$$

For the purpose of investigating the properties of the Heisenberg chain at wavelengths much larger than the lattice distance it is expedient to replace the Hamiltonian H by the continuum form

$$H = \frac{1}{2}\int dx \left(\frac{d\bar{S}}{dx}\right)^2 - h\int dx\,(S^z - 1), \tag{6.1}$$

obtained by assuming a slow variation of $\bar{S}_n$ over a lattice distance and expanding $\bar{S}_n \simeq \bar{S}(x)$. We have furthermore measured lengths in units of the lattice parameter a, energies in units of the exchange constant J, chosen the dimensionless magnetic field $\bar{h} = \bar{H}_0/J$ in the positive z direction, and subtracted the ground state energy. The unit length spin field $\bar{S}(x)$ now satisfies the Poisson bracket algebra

$$\{S^\alpha(x), S^\beta(y)\} = -\delta(x-y)\sum_\gamma \varepsilon^{\alpha\beta\gamma} S^\gamma(x), \tag{6.2}$$

and the equation of motion $d\bar{S}/dt = \{H, \bar{S}\}$ takes the form

$$\frac{d\bar{S}}{dt} = \bar{S}\times\frac{d^2\bar{S}}{dx^2} + \bar{S}\times\bar{h}, \quad \bar{h} = (0,0,h). \tag{6.3}$$

The above precessional equation of motion was first derived on phenomenological ground by Landau and Lifshitz[91] and later by Döring[92]. The approach by means of Poisson brackets is due to Mermin[93,94]. We emphasise that the continuum form (6.1) only correctly samples the long wavelength spin configurations of the Heisenberg chain. In the continuum limit "neighbouring" spins deviate only little with respect to one another, the spin field, however, "floats over all directions, as indicated in Fig.6.1, where we have shown an arbitrary spin configuration S(x).

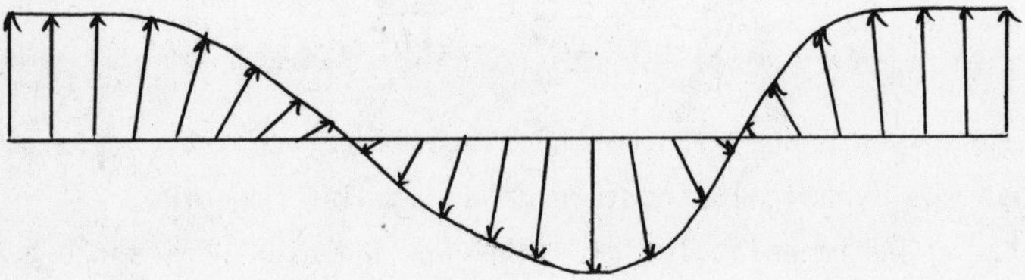

Fig.6.1 Arbitrary spin configuration with envelope shown

6.2 Hamiltonian Formulation - Constants of Motion

In order to exhibit the non linear character of the spin problem, caused by the precessional self coupling $S\times d^2S/dx^2$, within the framework of ordinary Hamiltonian dynamics[68], we introduce canonical variables. This will also enable us to identify easily all the constants of motion associated with the global symmetries of the hamiltonian.

Following Tjon and Wright[95] we define the canonical coordinate q(x) and the canonical momentum p(x) according to

$$S^x(x) = \sqrt{1-p(x)^2}\cos(q(x)) \tag{6.4a}$$

$$S^y(x) = \sqrt{1-p(x)^2}\sin(q(x)) \tag{6.4b}$$

$$S^z(x) = p(x), \tag{6.4c}$$

and it is readily shown that (6.2) implies the canonical Poisson brackets

$$\{p(x), q(y)\} = \delta(x-y) \tag{6.5a}$$

$$\{p(x), p(y)\} = 0 \tag{6.5b}$$

$$\{q(x), q(y)\} = 0 . \tag{6.5c}$$

Notice, however, that the canonical basis (6.4a-c) is not unique but depends on the choice of spin coordinate system. Since the spin Poisson bracket (6.2) in the form

$$\sum_n \frac{d\bar{S}(x)}{dp_n} \times \frac{d\bar{S}(y)}{dq_n} = \frac{a}{2}\,\delta(x-y)\,\bar{S}(x)$$

is manifestly covariant under rotations of the spin frame, the canonical representations are, however, related by canonical transformations[68].

In terms of the canonical variables p(x) and q(x) the spin Hamiltonian (6.1) takes the form

$$H = \frac{1}{2}\int dx\left(\frac{1}{1-p^2}\left(\frac{dp}{dx}\right)^2 + (1-p^2)\left(\frac{dq}{dx}\right)^2\right) - h\int dx\,(p-1)\,. \qquad (6.6)$$

The corresponding Hamiltonian equations of motion,

$$\frac{dq}{dt} = \{H,q\} = -\frac{1}{1-p^2}\frac{d^2p}{dx^2} - \frac{p}{(1-p^2)^2}\left(\frac{dp}{dx}\right)^2 - p\left(\frac{dq}{dx}\right)^2 - h \qquad (6.7a)$$

$$\frac{dp}{dt} = \{H,p\} = (1-p^2)\frac{d^2q}{dx^2} - 2p\,\frac{dp}{dx}\frac{dq}{dx}\,, \qquad (6.7b)$$

are, of course, equivalent to the field equation (6.3). In contrast to the Hamiltonian for a particle system H = T(p) + U(q), where T(p) is the kinetic energy and U(q) the potential energy, the form of the Hamiltonian (6.6) does not allow for a simple particle like interpretation and shows the strong intrinsic non linear character of the spin problem.

The total momentum Π is the generator of space translations and is defined by the Poisson bracket relation[68]

$$\frac{dF}{dx} = -\{\Pi, F\}\,, \qquad (6.9)$$

where F is an arbitrary function of the canonical variables p and q. For the spin density, in particular, we have

$$\frac{d\bar{S}}{dx} = -\{\Pi, \bar{S}\}\,. \qquad (6.10)$$

The Poisson bracket of Π with the Hamiltonian (6.1) for a "string" of length 2L,

$$\{\Pi, H\} = \left[\frac{1}{2}\left(\frac{d\bar{S}}{dx}\right)^2 - h\,S^z\right]_{x=-L}^{x=L}$$

vanishes provided we impose either the fixed boundary conditions $S^z \to 1$ and $dS/dx \to 0$ for $|x| = L \to \infty$ or the periodic boundary conditions $S^z(L) = S^z(-L)$ and $(dS/dx)_{x=L} = (dS/dx)_{x=-L}$. The Hamiltonian is thus translationally invariant and the total momentum Π a constant of motion, i.e., $\{\Pi, H\} = 0$. In the canonical basis the total momentum is given by the integrated density[95],

$$\Pi = \int dx\, (1-p) \frac{dq}{dx}, \tag{6.11}$$

where we, in order to ensure a vanishing total momentum in the ground state $S^z = p = 1$, have subtracted a total derivative. Correspondingly, in spin space by inserting (6.4a-c),

$$\Pi = \int dx \frac{S^x \frac{dS^y}{dx} - S^y \frac{dS^x}{dx}}{1 + S^z}. \tag{6.12}$$

Introducing a unit vector $\bar{n}$ in the z direction we can express the momentum density in the form

$$\pi(x) = \frac{\bar{n}\cdot \bar{S} \times \frac{d\bar{S}}{dx}}{1 + \bar{n}\cdot \bar{S}}$$

which shows that $\pi(x)$ depends explicitly on the uniform "gauge field" $\bar{n}$, characterising the spin frame. In contrast to the energy density $\mathcal{E}(x) = \frac{1}{2}(d\bar{S}/dx)^2 - \bar{h}\cdot\bar{S}$ which is "gauge independent", the momentum density is therefore not unambiguously defined, in other words, we cannot localise the momentum. The total momentum, however, is independent of the gauge since $\int dx p dq/dx$ is canonically invariant under symmetry transformations[68]. An analogous situation is, incidentally, encountered in general relativity with regard to the localisation of the energy.

The total spin or angular momentum $\bar{M}$ is the generator of rotations in spin space. We have the Poisson bracket relation

$$\{M^\alpha, S^\beta(x)\} = -\sum_\gamma \varepsilon^{\alpha\beta\gamma} S^\gamma(x), \tag{6.13}$$

and it follows from (6.2) that

$$M^x = \int dx\, S^x(x) \tag{6.14a}$$

$$M^y = \int dx\, S^y(x) \tag{6.14b}$$

$$M^z = \int dx\, (S^z(x) - 1)\,, \tag{6.14c}$$

where we have subtracted the value in the ground state. In the canonical basis (6.4a-c) the angular momentum takes the form

$$M^x = \int dx \sqrt{1-p^2}\, \cos(q) \tag{6.15a}$$

$$M^y = \int dx \sqrt{1-p^2}\, \sin(q) \tag{6.15b}$$

$$M^z = \int dx\, (p-1)\,. \tag{6.15c}$$

In the absence of the magnetic field $\bar{h}$ = (0.0.h) the Hamiltonian (6.1) is invariant under rotation in spin space and all three components of the total angular momentum are constants of motion, i.e., $\{\bar{M},H\} = 0$. In the presence of the field $\bar{h}$ the Hamiltonian remains invariant under rotations about the z axis. Consequently, $\{M^z,H\} = 0$ and only M^z is a constant of motion.

6.3 Permanent Profile Solutions - General Discussion

Prior to discussing the general dynamical solutions of the continuous Heisenberg chain (6.1), it is instructive to consider a class of special solutions which can be derived by quadrature, namely spin configurations propagating with a permanent profile.

We search, in other words, for solutions of the form $S(x,t) = S(x-vt)$, where v is the phase velocity of the permanent profile. In terms of the canonical equations of motion (7.7a-b) we obtain, inserting $q = q(x-vt)$ and $p = p(x-vt)$,

$$-v\frac{dq}{dx} = -\frac{1}{1-p^2}\frac{d^2p}{dx^2} - \frac{p}{(1-p^2)^2}\left(\frac{dp}{dx}\right)^2 - p\left(\frac{dq}{dx}\right)^2 - h$$

$$-v\frac{dp}{dx} = (1-p^2)\frac{d^2q}{dx^2} - 2p\frac{dp}{dx}\frac{dq}{dx}\,.$$

The second equation is readily integrated once,

$$\frac{dq}{dx} = v\,\frac{p_0 - p}{1-p^2}\,. \tag{6.16}$$

Substituting (6.16) in the first equation we obtain a second order differential equation for p which integrated once yields the first order non linear equation

$$\left(\frac{dp}{dx}\right)^2 = F(p) = 2hp(p^2-1) - v^2(1+p_0-2p_0p) - p_1(p^2-1), \tag{6.17}$$

where p_0 and p_1 are constants of integration to be determined by the choice of permanent profile solution. The general solution of (6.17) is given by an elliptic function[96], however, for the present purposes an elementary discussion suffices.

In order to obtain a solution of (6.17) we must choose p_0 and p_1 such that the cubic polynomial F(p) is positive for p in the admissible range $-1 < p < 1$. We distinguish three cases.

I) F(p) has a double root p_A and is negative everywhere else in the interval $-1 < p < 1$. In this case p is tied to the value p_A and the equation (6.16) for q has the solution $q = k(x-vt) + q_0$, where $k = (p_0-p_A)/(1-p_A^2)$ and q_0 is an integration constant. In spin space $S^z = p$, $S^+ = (1-p^2)^{1/2}\exp(iq)$ and the above solution corresponds to the propagation of a spin wave with wavenumber k, frequency vk, and a constant projection on the z axis. In Fig.6.2 we have plotted p(xt) and q(xt), modulus 2π, in the case of spin wave motion, and in Fig.6.3 the corresponding form of F(p).

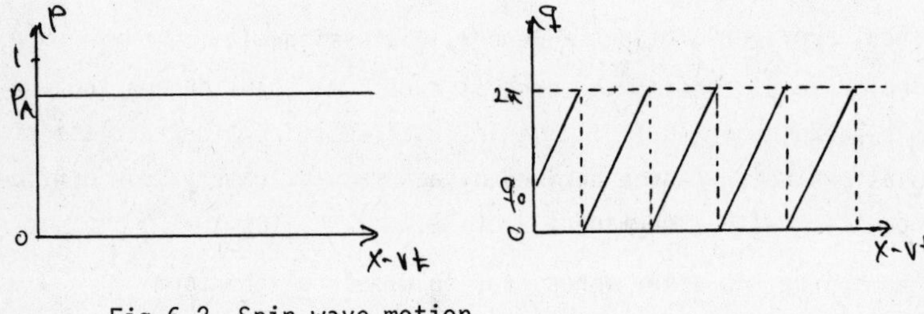

Fig.6.2 Spin wave motion

II) F(p) has a single root p_A in the interval $-1 < p < 1$ and a double root at one of the end points, say $p = 1$. Furthermore, F(p) is positive in the interval $p_A < p < 1$ and negative everywhere else in $-1 < p < 1$. In this case (6.17) has a turning point[97] at $p = p_A$ and a degenerate turning point at $p = 1$. The motion of p is restricted to the interval $p_A < p < 1$. For

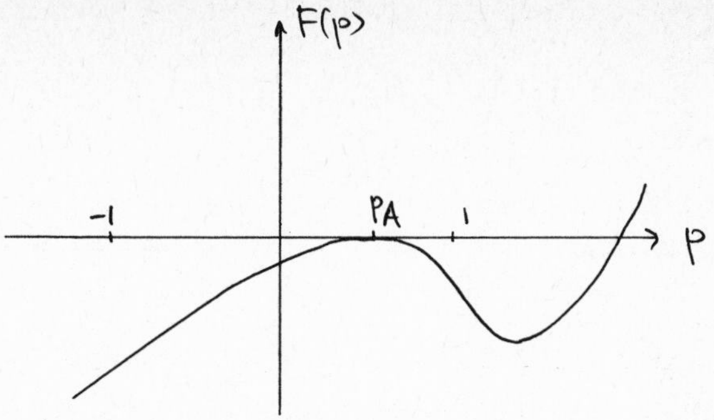

Fig.6.3 The form of F(p) in the case of spin wave motion

$|x| \to \infty$ p approaches asymptotically the ground state value +1. In order to ensure a solution of (6.16) we choose $p_o = 1$, i.e., $q = v\int_{-\infty}^{x-vt} dy(1+p(y))^{-1}$. For (dp/dx) to vanish for $|x| \to \infty$ we must furthermore choose $p_1 = v^2 + 2h$. In spin space the above solution corresponds to the propagation of a solitary wave with phase velocity v. In Fig.6.4 we have sketched the solution p(xt) and q(xt), modulus 2π, for a solitary wave motion. In Fig.6.5 we have shown the appropriate form of F(p).

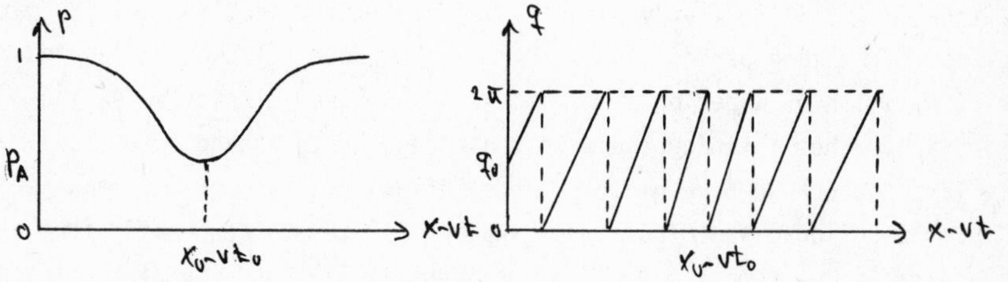

Fig.6.4 Solitary wave motion

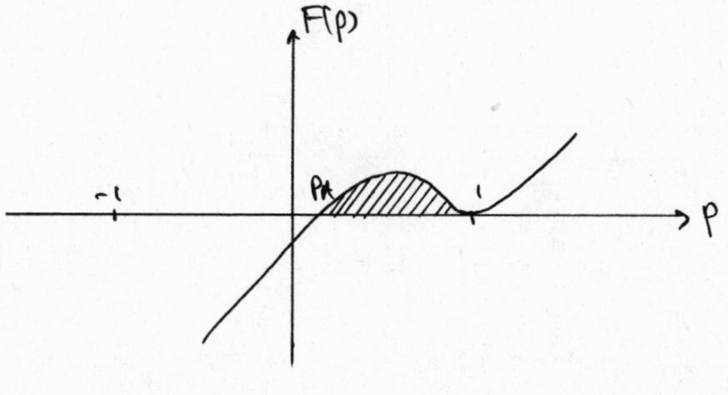

Fig.6.5 The form of F(p) in the case of solitary wave motion

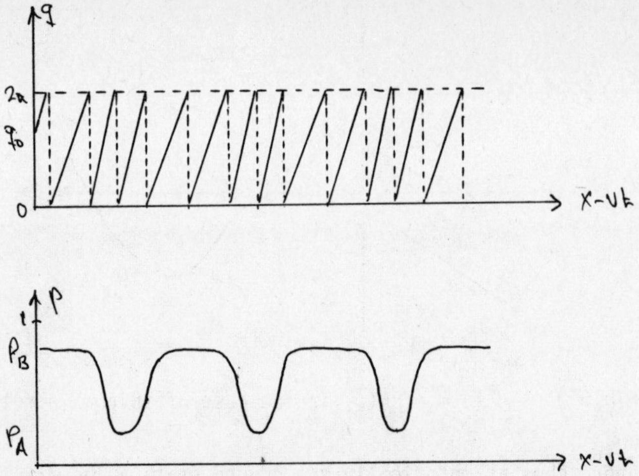

Fig.6.6 Periodic wave train

III) F(p) has two distinct roots at p_A and p_B in the interval $-1 < p < 1$, is positive for $p_A < p < p_B$ and negative everywhere else. In this case (6.17) has turning points at both p_A and p_B. The solution is restricted to the interval $p_A < p < p_B$ and is periodic with period $2\int_{p_A}^{p_B} dp1/(F(p))^{1/2}$. The solution of (6.16), $q = v\int_{-\infty}^{x-vt} dy(p_0 - p(y))/(1-p(y))$ is also, modulus 2π, periodic with the same period. In spin space the above solution corresponds to the propagation of a periodic wave train with phase velocity v. As the period approaches infinity the wave train reduces to a single solitary wave. In Fig. 6.6 we have plotted the solution p(xt) and q(xt), modulus 2π, for a periodic wave train, and in Fig. 6.7 the corresponding form of F(p). Finally, we remark that the case where (6.17) has a turning point at

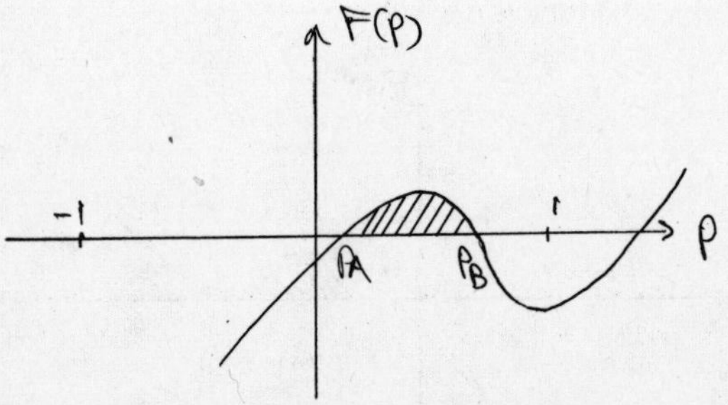

Fig. 6.7 The form of F(p) is the case of a periodic wave train

p_A and a degenerate turning point at p_B, $|p_B| < 1$, is a variant of case II, corresponding to a rotation of the spin frame, i.e., a solitary wave with p approaching the ground state value p_B for $|x| \to \infty$. Also, for certain values of p_0 and p_1 the periodic wave train in case III degenerates to a spin wave in a rotated frame.

6.4 Spin Waves - Solitary Waves - Periodic Wave Trains

As discussed in the previous section, the permanent profile spin wave solutions are characterised by having a constant z component. From the equation of motion (6.3) we thus obtain, introducing $S^{\pm} = S^x \pm iS^y$, $dS^+/dt = iS^z d^2S^+/dx^2 - iS^+h$ with the constraint $S^- d^2S^+/dx^2 = S^+ d^2S^-/dx^2$. Imposing, furthermore, the length condition $S^+S^- + (S^z)^2 = 1$, we infer in accordance with Laksmanan et al.[98] the spin wave solution

$$S^+(x,t) = \sqrt{1-(S^z)^2}\, \exp(i(kx-\omega t+\phi)), \tag{6.18}$$

specified by the amplitude S^z, the wavenumber k, the field h, and the phase ϕ. The frequency is given by the dispersion law

$$\omega = S^z k^2 + h. \tag{6.19}$$

In contrast to a quantum mechanical spin wave[11,12], the classical counterpart (6.18) forms a band even for a fixed value of k, as shown in Fig.6.8, where we have plotted ω versus k for $-1 < S^z < 1$.

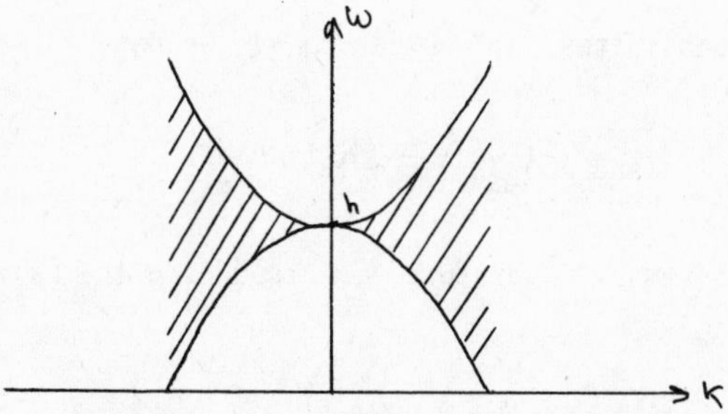

Fig.6.8 Plot of $\omega = S^z k^2 + h$. The shaded area indicates the spin wave continuum (arbitrary units)

We also notice that for $h = -S^z k^2$, i.e., $-k^2 < h < k^2$, we have a band of static spin waves.

From the Hamiltonian (6.1) the energy density of a spin configuration is given by $\frac{1}{2}(dS/dx)^2 - h(S^z-1)$. For the spin wave solution (6.18), in particular, we obtain by insertion the constant energy density

$$\varepsilon = \tfrac{1}{2}k^2(1-(S^z)^2) - h(S^z-1),$$

i.e., the total energy of the non localised spin wave is infinite. We also note that the largest energy density is attained by the band of static spin waves for $h = -S^z k^2$.

We emphasise that the spin wave spectrum discussed here is an exact solution of the equation of motion (6.3). By considering small deviations from the aligned ground state configuration $S^z = 1$ we obtain a linearised spectrum with dispersion law $\omega = k^2 + h$. It is maybe interesting to notice that the sole effect of the local mode coupling $\bar{S} \times d^2S/dx^2$, which in wavenumber space gives rise to a non local interaction, is to change the stiffness coefficient, i.e., the coefficient of k^2, from unity to S^z.

In order to obtain the permanent profile solitary wave solution first derived by Nakamura and Sasada[99] (see also refs.98 and 95) we choose, as discussed in the previous section, $p_0 = 1$ and $p_1 = v^2+2h$ in the canonical equations of motion (6.16) and (6.17), i.e.,

$$\frac{dq}{dx} = \frac{v}{1+p} \quad \text{and} \quad \left(\frac{dp}{dx}\right)^2 = (p-1)^2(2h(p+1) - v^2).$$

Introducing polar coordinates $p = \cos\theta$ and $q = \phi$ we have

$$\frac{d\phi}{dx} = \frac{v}{1+\cos\theta} \quad \text{and} \quad \left(\frac{d\theta}{dx}\right)^2 = 2h(1-\cos\theta) - v^2\left(\frac{1-\cos\theta}{1+\cos\theta}\right)$$

which are readily solved by quadrature[95], introducing the half angle $\theta/2$, i.e.,

$$\left(\frac{d\theta}{dx}\right)^2 = 4h\left(\cos^2\left(\frac{\theta}{2}\right) - \frac{v^2}{4h}\right)\tan^2\left(\frac{\theta}{2}\right)$$

Hence, we obtain the solitary wave solution

$$\cos\theta = 1 - \frac{A}{\cosh^2\left(\frac{x-vt-x_0}{\Gamma}\right)} \tag{6.20a}$$

$$\phi = \phi_0 + \tfrac{1}{2}v(x-vt-x_0) + \tan^{-1}\left(\frac{2}{v\Gamma}\tanh\left(\frac{x-vt-x_0}{\Gamma}\right)\right), \tag{6.20b}$$

where we have introduced the amplitude

$$A = 2 - \frac{v^2}{2h} \tag{6.21}$$

and the width

$$\Gamma = \frac{1}{\sqrt{h - \left(\frac{v}{2}\right)^2}} . \tag{6.22}$$

The center of mass x_0 and the phase ϕ_0 are determined by the initial conditions.

The solitary wave solution (6.20a-b) is characterised by the four parameters A, Γ, x_0, and ϕ_0. In contrast to the spin wave velocity ω/k, the phase velocity of the solitary wave is restricted to the interval $-2h^{\frac{1}{2}} < v < 2h^{\frac{1}{2}}$. In the low velocity limit the amplitude A attains its maximum value 2, and the width Γ its minimum value $1/h^{1/2}$. On the other hand, when the velocity approaches its maximum values $\pm 2h^{1/2}$, the amplitude A vanishes and the width Γ becomes infinite, i.e., the solitary wave disappears. In the limit h = 0 the velocity is at its maximum values and a permanent profile solitary wave solution does not exist. In Figs. 6.9 and 6.10 we have depicted the z component of the solitary wave in the two cases. The phase of the solitary wave develops essentially linearly with x - vt, modulus 2π, but picks up a positive phase shift $\Delta\phi$ given by

$$\Delta\phi = 2\tan^{-1}\left(\frac{2}{v\Gamma}\right) \tag{6.23}$$

across a region of width Γ about the center of mass x_0, as shown in Fig. 6.11. As the velocity approaches zero, the phase shift $\Delta\phi$ attains its maximum value π; for $v = \pm 2h^{\frac{1}{2}}$ the phase shift vanishes together with the solitary wave. In Fig.6.12 we have plotted $\Delta\phi$ as a function of v. In order to illustrate the phase shift effect it is instructive to plot the projection of the spin field of a solitary wave onto the XY plane. In Figs.6.13, 6.14, and 6.15 we have shown the three cases of a small amplitude, a half and a full amplitude solitary wave.

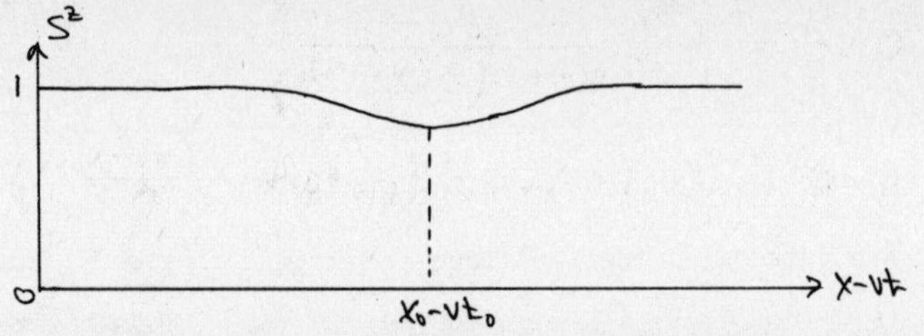

Fig.6.9 Small amplitude-large width-large velocity solitary wave (arbitrary units)

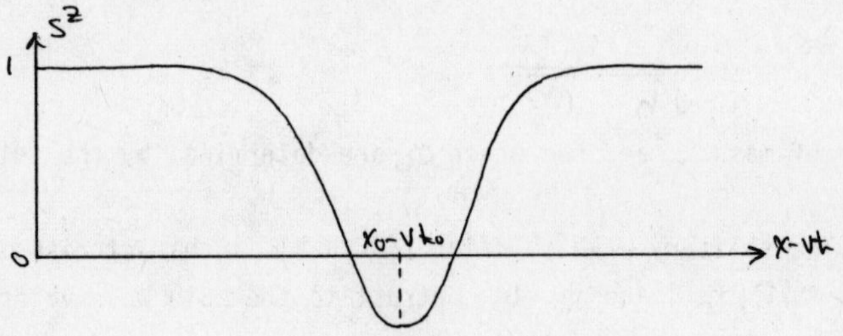

Fig.6.10 Large amplitude-small width-small velocity solitary wave (arbitrary units)

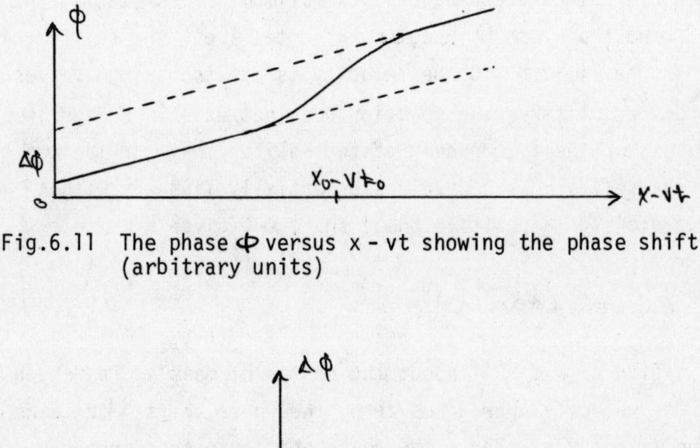

Fig.6.11 The phase ϕ versus x - vt showing the phase shift $\Delta\phi$ (arbitrary units)

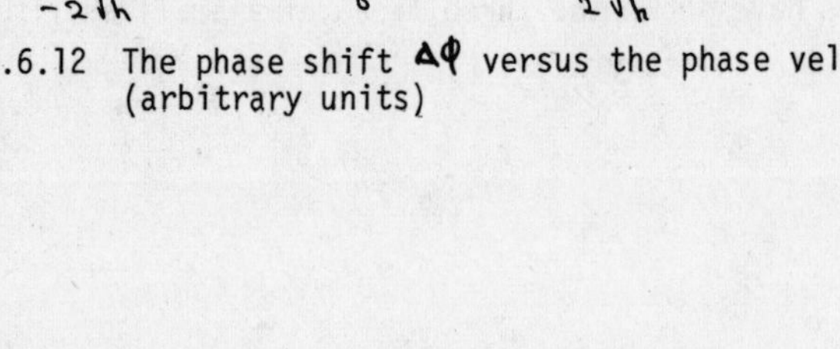

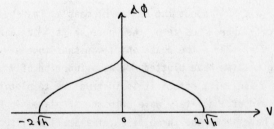

Fig.6.12 The phase shift $\Delta\phi$ versus the phase velocity v (arbitrary units)

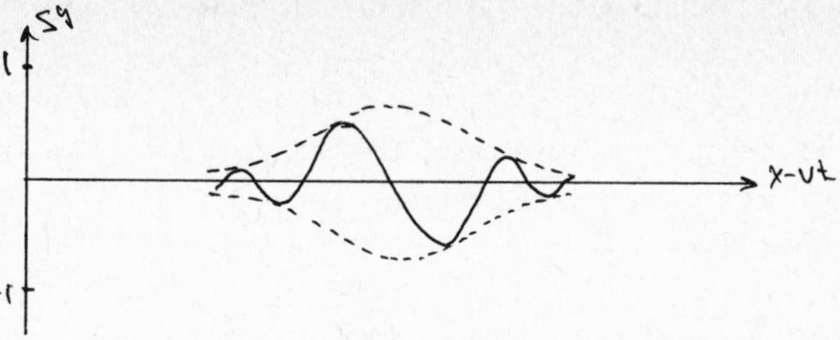

Fig.6.13 The transverse component for a small amplitude solitary wave; the dashed line indicates the envelope (arbitrary units)

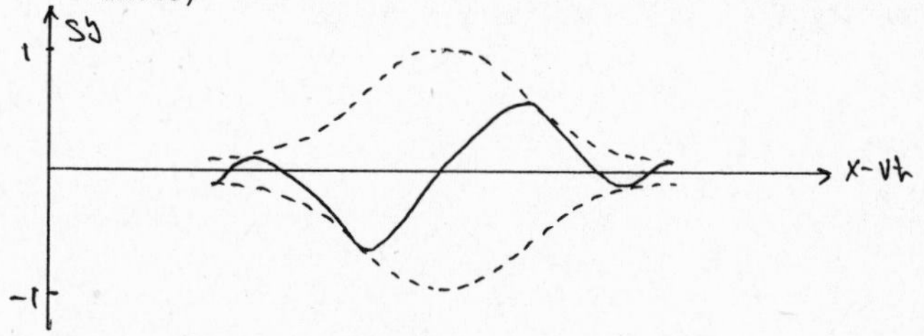

Fig.6.14 The transverse component for a half amplitude solitary wave; the dashed line indicates the envelope (arbitrary units)

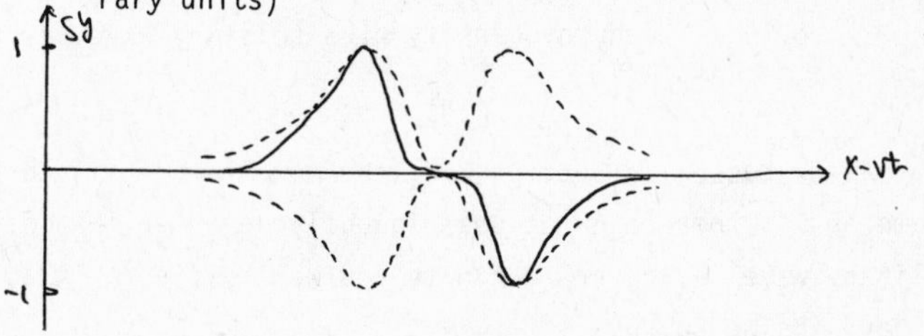

Fig.6.15 The transverse component for a full amplitude solitary wave; the dashed line indicates the envelope (arbitrary units)

From Eq.(6.6) the energy density is in polar coordinates $p = \cos\theta$, $q = \phi$ given by

$$\varepsilon(x,t) = \frac{1}{2}\left(\left(\frac{d\theta}{dx}\right)^2 + \sin^2\theta\left(\frac{d\phi}{dx}\right)^2\right) - h(\cos\theta - 1) .$$

Substituting the solitary wave solution (6.20a-b) and introducing the amplitude A (6.21) and width Γ (6.22), notice that $h = 2/A\Gamma^2$, we obtain

$$\varepsilon(xt) = \frac{\frac{4}{\Gamma^2}}{\cosh^2\left(\frac{x-vt-x_0}{\Gamma}\right)} . \tag{6.24}$$

In contrast to the spin wave case, the energy density of a solitary wave is not uniform but peaked at the center-of-mass position. The finite total energy of a solitary wave is

$$E = \frac{8}{\Gamma} = 8\sqrt{h-\left(\frac{v}{2}\right)^2} , \tag{6.25}$$

i.e., inversely proportional to the width. In Fig.6.16 we have plotted the energy density $\varepsilon(xt)$ of a solitary wave.

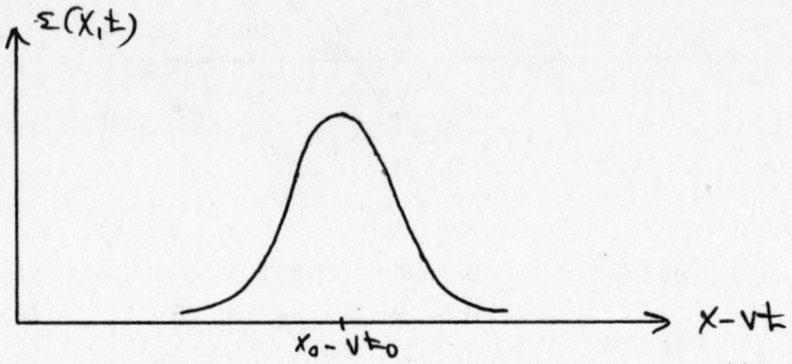

Fig.6.16 The energy density of a solitary wave (arbitrary (units)

As discussed in Section 6.2, the momentum density depends on the spin frame and is therefore not unambiguously defined. The total momentum of a solitary wave is, however, finite and well-defined. In polar coordinates we have, using Eq.(6.11)

$$\Pi = \int dx\, (1-\cos\theta)\frac{d\phi}{dx} .$$

Inserting the solitary wave solution (6.20a-b) we obtain by quadruture

$$\Pi = 4\sin^{-1}\left(\sqrt{\frac{A}{2}}\right) - 4\sin^{-1}\left(\sqrt{1-\frac{v^2}{4h}}\right) . \tag{6.26}$$

The total momentum of a solitary wave is restricted to lie in the interval $-2\pi < \Pi < 2\pi$. The momentum assumes its maximum value $|\Pi| = 2\pi$ for a full amplitude solitary wave, the value $|\Pi| = \pi$ for a half amplitude wave, and vanishes together with the solitary wave. The momentum-velocity relationship for a solitary wave is peculiar as seen by solving (6.26) for v,

$$|v| = 2\sqrt{h}\,\cos\left(\frac{\tilde{\Pi}}{4}\right) .$$

The velocity attains its maximum value $2h^{\frac{1}{2}}$ for vanishing momentum, and vice versa. This bizarre relationship is depicted in Fig.6.17.

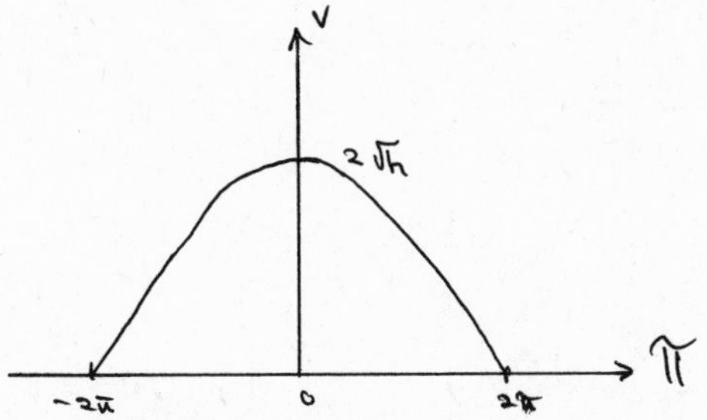

Fig.6.17 The velocity-momentum relationship for a solitary wave (arbitrary units)

The angular momentum density is given by the spin field itself. In polar coordinates we obtain from Eq.(6.15c) the z component of the total angular momentum,

$$M^z = \int dx\,(\rho - 1) = \int dx\,(\cos\theta - 1) .$$

Substituting the solitary wave solution (6.20a-b) we have by quadrature

$$M^z = -\frac{4}{h\Gamma} = -4\,\frac{\sqrt{h - \left(\frac{v}{2}\right)^2}}{h} . \qquad (6.27)$$

We notice that the total angular momentum of the solitary wave is inversely proportional to its width.

Instead of characterising the solitary wave by means of the amplitude A and the width Γ we can use the constants of motion E, $\tilde{\Pi}$ and M^z. The energy, momentum, and angular momentum are related by the dispersion law

$$E = \frac{16}{|M^z|}\sin^2\left(\frac{\tilde{\Pi}}{4}\right) + h|M^z| = \frac{32}{|M^z|}\sin^2\left(\frac{\tilde{\Pi}}{4}\right) , \qquad (6.28)$$

obtained from Eqs. (6.25), (6.26), and (6.27) by eliminating A and Γ, using $h = 2/A\Gamma^2$. In the low momentum limit $\Pi \ll 1$, $E = \Pi^2/|M^z|$, and the solitary wave propagates as a free particle with an effective mass $|M^z|/2$. This particle analogy, however, breaks down for larger values of Π, in particular as we approach the maximum values $\pm 2\pi$. In Fig.6.18 we have shown the energy-momentum relationship for a solitary wave for different values of the angular momentum M^z.

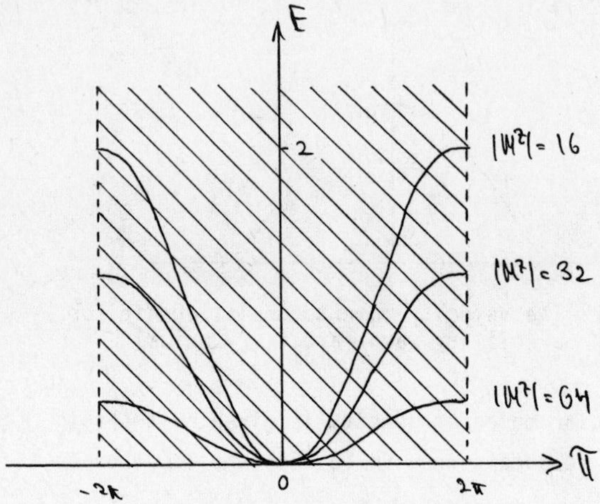

Fig.6.18 The dispersion law for a solitary wave for $|M^z|$ = 16 , 32 , 64. The shaded area indicates the solitary wave band (arbitrary units)

We remark that similar to the spin wave spectrum (see Fig.6.8) the solitary waves form a band since M^z ranges continuously over negative values, as shown in Fig.6.8.

The permanent profile wave train solutions are, as mentioned in Section 6.3 given by an elliptic function[97] and we shall not discuss the analytic structure here. In the light of the above discussion of the solitary wave solutions we are, however, able to draw some simple conclusions. We can visualise the wave train as a periodic lattice of solitary waves with a periodic energy density localised at the center-of-mass positions. The phase ϕ increases linearly with x-vt but picks up a phase shift $\Delta\phi$ each time we traverse a solitary wave peak. In the limit where the period becomes large the ground state is nearly established in the regions between the solitary waves. In Fig.6.19 we have depicted the longitudinal and transverse

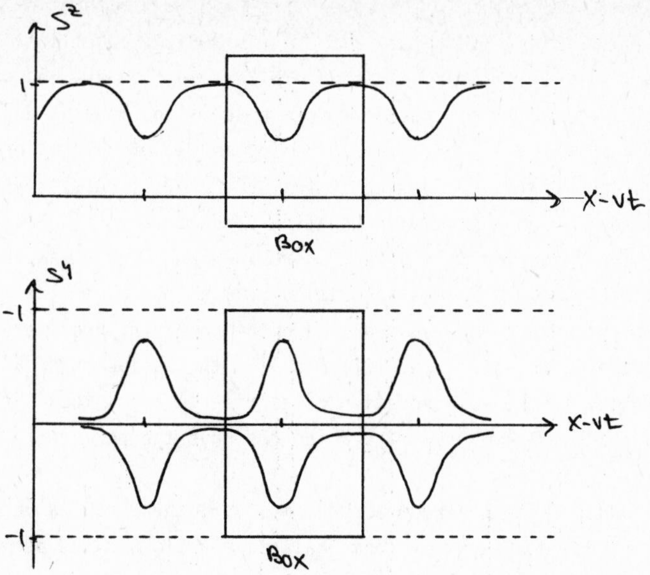

Fig.6.19 Periodic wave train solution; the dashed line indicates the envelope (arbitrary units)

spin components for a periodic wave train. In the case of periodic boundary conditions in a finite box it is interesting to notice that the periodic wave train solution with period L essentially corresponds to inserting a single solitary wave in a box of size L, as shown in Fig.6.19.

We conclude this section with some general remarks concerning permanent profile solutions (see also refs. 89 and 100). Let us first consider the ubiquitous wave equation[71,101]

$$\frac{d^2\phi}{dt^2} - v^2\frac{d^2\phi}{dx^2} = 0$$

which has the general permanent profile solution $f(x \pm vt)$, where f is an arbitrary function. Since the equation is linear we can, of course, express the general solution Φ as a superposition of plane waves

$$\phi(x,t) = \int \frac{dk}{2\pi} A_k e^{ik(x-vt)} + \int \frac{dk}{2\pi} B_k e^{ik(x+vt)}$$

Three basic mechanisms will in general destroy the permanent profile solution, namely I) dissipation, II) dispersion, and III) non linearity. I) The inclusion of a dissipative term, say $-\gamma d\phi/dx$, in the wave equation gives rise to a damping of the permanent profile solution. This is an irreversible effect which cannot be balanced by any amount of dispersion or non linearity, and we shall not consider it further here. II) A dispersive term, say $-\alpha d^4\phi/dx^4$, added to the wave equation leads to a non linear frequency-wavenumber relationship $\omega = v(k)k$, where in the present example $v(k) = (v^2+\alpha k^2)^{\frac{1}{2}}$. The phase velocities $v(k)$ of the different plane wave component depend on the wavenumber k and the permanent profile solution breaks up due to interference. III) Finally, the inclusion of a non linear term, say $\lambda\phi^2$, in the wave equation gives rise to a non local interaction in wavenumber space, $\lambda\int\frac{dp}{2\pi}\phi_{k-p}\phi_p$, which destroys the permanent profile solution.

Under certain circumstances, however, the dynamical effects of the dispersive and non linear terms balance exactly and the permanent profile solutions are stable. This is in particular the case for the precessional equation of motion (6.3), $d\bar{S}/dt = \bar{S}\times d^2\bar{S}/dx^2 + \bar{S}\times\bar{h}$, which has a quadratic dispersion and a quadratic non linearity.

6.5 The Lax Representation

In this section we begin the general analysis of the dynamics of the continuous Heisenberg chain (6.1). Since the magnetic field term in the equation of motion (6.3) can be absorbed by a transformation to a rotating spin frame $S^+ \rightarrow S^+\exp(-iht)$, we consider in the following, without loss of generality,

$$\frac{d\bar{S}}{dt} = \bar{S}\times\frac{d^2\bar{S}}{dx^2} \tag{6.29}$$

Following Takhtajan[102] we imbed the equation of motion (6.29) in the Pauli matrix basis

$$\sigma^x = \begin{pmatrix} & 1 \\ 1 & \end{pmatrix}, \quad \sigma^y = \begin{pmatrix} & -i \\ i & \end{pmatrix}, \quad \sigma^z = \begin{pmatrix} 1 & \\ & -1 \end{pmatrix}$$

$$\sigma^\alpha\sigma^\beta = \delta^{\alpha\beta} + i\sum_\gamma \varepsilon^{\alpha\beta\gamma}\sigma^\gamma .$$

Introducing the traceless self-adjoint spin matrix

$$S = \sum_{\alpha} S^{\alpha}\sigma^{\alpha} = \begin{vmatrix} S^z & S^- \\ S^+ & -S^z \end{vmatrix} , \tag{6.30}$$

where the length condition $\bar{S}^2 = 1$, furthermore, implies that $S^2 = I$ and $\det S = 1$ (det denotes the determinant), the equation of motion (6.29) takes the matrix form

$$\frac{dS}{dt} = -\frac{i}{2}\left[S, \frac{d^2S}{dx^2}\right] . \tag{6.31}$$

In order to "monitor" the instantaneous spin configuration S(xt) we consider the associated eigenvalue problem $iL\Psi = \lambda\Psi$, where

$$L = S(xt)\frac{d}{dx} = \begin{vmatrix} S^z(xt) & S^-(xt) \\ S^+(xt) & -S^z(xt) \end{vmatrix} \frac{d}{dx}$$

is a linear operator acting in a space of x dependent matrices . The operator L depends parametrically on the spin field S(xt) and is therefore in general time dependent. It has, however, been shown by Takhtajan[102] that the non linear evolution equation (6.31) for the spin field induces a similarity transformation of L, $L(t) = U(t)L(0)U^{-1}(t)$. This is a crucial observation which immediately implies that the spectrum $\{\lambda\}$ of the operator L(t) is time independent, i.e., a constant of motion. The eigenfunction Ψ picks up all the time dependence and evolves according to the similarity transformation U(t), $\Psi(t) = U(t)\Psi(0)$. The time evolution of the operator U(t) is governed by $i\frac{dU}{dt} = MU$, $U(0) = I$, where the auxiliary operator $M = 2Sd^2/dx^2 + (dS/dx)d/dx$, and it follows that L and Ψ satisfy the equations of motion $\frac{dL}{dt} = i[L,M]$ and $i\frac{d\Psi}{dt} = M\Psi$.

The importance of the Lax[102] representation,

$$L = S\frac{d}{dx} , \; M = 2S\frac{d^2}{dx^2} + \frac{dS}{dx}\frac{d}{dx} , \; \frac{dL}{dt} = i[L,M] \tag{6.32}$$

lies in the fact that it replaces the in general intractable problem of solving the non linear equation of motion (6.31) for S(xt) directly by the solution of two linear operator problems,

$$iS\frac{d\psi}{dx} = \lambda\psi, \quad i\frac{d\psi}{dt} = 2S\frac{d^2\psi}{dx^2} + \frac{dS}{dx}\frac{d\psi}{dx}. \qquad (6.33)$$

Since there exists a variety of mathematical techniques for linear problems but practically none for non linear ones, this represents a major simplification (see, for instance) refs.89 and 90).

A representation of the kind (6.32) was first introduced for the non linear Kortweg-deVries equation, encountered in shallow water wave theory[100],

$$\frac{du}{dt} - 6u\frac{du}{dx} + \frac{d^3u}{dx^3} = 0,$$

in the pioneering work by Gardner, Green, Kruskal, and Miura[103], and later refined by Lax[104], hence the name. Since that time Lax representations have been found for a whole class of one dimensional non linear evolution equations, comprising among others the non linear Schrödinger equation[105]

$$i\frac{d\psi}{dt} + \frac{d^2\psi}{dx^2} + \kappa|\psi|^2\psi = 0$$

and the ubiquitous Sine Gordon equation[106]

$$\frac{d^2u}{dt^2} - \frac{d^2u}{dx^2} + \sin(u) = 0.$$

6.6 The associated Eigenvalue Problem - General Discussion

Owing to the Lax representation (6.32) the spectrum $\{\lambda\}$ of the linear eigenvalue problem $iL\psi = \lambda\psi$ is time independent and is therefore related to the constants of motion of the continuous Heisenberg chain. Here we discuss the eigenvalue problem in some detail, following and expanding the work in ref.102. Encumbering mathematical details are, however, deferred to the following section.

By means of the idempotent property $S^2 = I$ we can express the eigenvalue equation $iL\psi = \lambda\psi$ in the more convenient form

$$i\frac{d\psi}{dx} = \lambda S\psi. \qquad (6.34)$$

Let us first discuss some general properties. Since (6.34) is of first order in d/dx, two distinct solutions ψ_1 and ψ_2 are related by a constant matrix, i.e., $\psi_1 = \psi_2 A$. This follows easily by taking a derivative of $\psi_2^{-1}\psi_1$ using (6.34). The property TrS = 0, furthermore, implies that the determinant of a solution, $\det\psi$, is a constant. Expressing $\det\psi$ in the form $\exp(\mathrm{Tr}\log\psi)$ and applying $i\frac{d}{dx}$ we obtain, using (6.34) and the cyclic permutability of operators under the trace operation, $i\frac{d}{dt}\det\psi = \det\psi\,\mathrm{Tr}(\lambda S)$ with the solution $\det\psi(x)/\det\psi(x_0) = -i\lambda\int_{x_0}^{x} dy\,\mathrm{Tr}S(y)$, or since TrS = 0, $\det\psi$ = const. Finally, noticing that $S^* = -\sigma^y S\sigma^y$, we conclude that if ψ is a solution of (6.34) with eigenvalue λ then $\sigma^y\psi^*\sigma^y$ is a solution with the complex conjugate eigenvalue λ^*. Summarising,

$$\psi_1 \text{ and } \psi_2 \text{ solutions then } \psi_2 = \psi_1 A, \text{ A constant} \tag{6.35a}$$

$$\det\psi = \text{constant} \tag{6.35b}$$

$$\psi \text{ solution for } \lambda \text{ then } \sigma^y\psi^*\sigma^y \text{ solution for } \lambda^*. \tag{6.35c}$$

Introducing an ordering procedure in x space it is easy to derive a formal solution of (6.34). In analogy with the form of the evolution operator in the interaction representation[24,25] we obtain

$$\psi(x) = \left(\exp\left(-i\lambda\int_{x_0}^{x} S(y)dy\right)\right)_+ \psi(x_0), \tag{6.36}$$

where $(..)_+$ indicates an ordering in x space, for instance, $(S(x)S(y))_+ = S(x)S(y)$ for $x > y$ and $= S(y)S(x)$ for $x < y$.

In order to render the eigenvalue equation (6.34) precise, we must choose appropriate boundary conditions for the spin field S(x). There are two possibilities, namely either fixed boundary conditions at infinity, i.e., $S(x)\to\sigma^z$ for $|x|\to\infty$; or periodic boundary conditions in a box of size L, i.e., S(x+L) = S(x). We notice that of the special permanent profile solutions discussed in Section 6.4, the solitary waves correspond to fixed boundary conditions at infinity, whereas the spin waves and the periodic wave trains are only compatible with periodic boundary conditions; the spin wavenumber k then takes discrete values $k = 2\pi n/L$, n integer, and the period of the wave train is equal to the box size L. It turns out that the eigenvalue equation (6.34) is most easily discussed if we impose fixed boundary conditions at infinity for the "potential" S(x), i.e., $S(x)\to\sigma^z$ for $|x|\to\infty$. This choice, of course, corresponds to taking the infinite volume limit at the outset. The case of periodic boundary conditions $S(x+L) = S(x)$, which is equivalent to treating the system in a finite box of length L, presents technical difficulties and will not be considered in the present context.

Since $S(x) \to \sigma^z$ for $|x| \to \infty$ it follows that the solutions of the eigenvalue equation (6.34) for $|x| \to \infty$ have the general form $\Psi_\infty(x) = \exp(-i\lambda\sigma^z x)A$, where A is a constant matrix. It is particularly convenient to introduce the two Jost solutions $F(x\lambda)$ and $G(x\lambda)$ determined by the boundary conditions A = I for $x \to \infty$ and A = I for $x \to -\infty$ respectively, i.e.

$$F(x\lambda) \to \exp(-i\lambda\sigma^z x) \quad \text{for } x \to \infty \tag{6.37a}$$

$$G(x\lambda) \to \exp(-i\lambda\sigma^z x) \quad \text{for } x \to -\infty . \tag{6.37b}$$

In terms of the general formal solution (6.36) the Jost functions are given by

$$F(x\lambda) = \exp(-i\lambda\sigma^z x)\left(\exp\left(+i\int_x^\infty (S(y)-\sigma^z)dy\right)\right)_+ \tag{6.38a}$$

$$G(x\lambda) = \exp(-i\lambda\sigma^z x)\left(\exp\left(-i\int_{-\infty}^x (S(y)-\sigma^z)dy\right)\right)_+ , \tag{6.38b}$$

where we have used the space ordering property in order to isolate the factor $\exp(-i\lambda\sigma^z x)$. In Section 6.7 we discuss the spectral properties of the Jost functions and show that they satisfy the representations

$$F(x\lambda) = \exp(-i\lambda\sigma^z x) + \lambda\int_x^\infty K(xy)\exp(-i\lambda\sigma^z y)dy \tag{6.39a}$$

$$G(x\lambda) = \exp(-i\lambda\sigma^z x) + \lambda\int_{-\infty}^x N(xy)\exp(-i\lambda\sigma^z y)dy , \tag{6.39b}$$

where the kernels $K(xy)$ and $N(xy)$ depend functionally on the "potential" $S(x)$ but are independent of the eigenvalue λ. The restriction of the integration ranges in the Jost representations (6.39a-b) immediately imply an analytic continuation into the complex λ plane. In terms of the matrix elements we have

$$\begin{vmatrix} F_{11}(x\lambda) & F_{12}(x\lambda) \\ F_{21}(x\lambda) & F_{22}(x\lambda) \end{vmatrix} = \begin{vmatrix} e^{-i\lambda x} & \\ & e^{i\lambda x} \end{vmatrix} + \lambda \begin{vmatrix} \int_x^\infty K_{11}(xy)e^{-i\lambda y}dy & \int_x^\infty K_{12}(xy)e^{i\lambda y}dy \\ \int_x^\infty K_{21}(xy)e^{-i\lambda y}dy & \int_x^\infty K_{22}(xy)e^{i\lambda y}dy \end{vmatrix}$$

$$\begin{vmatrix} G_{11}(x\lambda) & G_{12}(x\lambda) \\ G_{21}(x\lambda) & G_{22}(x\lambda) \end{vmatrix} = \begin{vmatrix} e^{-i\lambda x} & \\ & e^{i\lambda x} \end{vmatrix} + \lambda \begin{vmatrix} \int_{-\infty}^x N_{11}(xy)e^{-i\lambda y}dy & \int_{-\infty}^x N_{12}(xy)e^{-i\lambda y}dy \\ \int_{-\infty}^x N_{21}(xy)e^{-i\lambda y}dy & \int_{-\infty}^x N_{22}(xy)e^{-i\lambda y}dy \end{vmatrix}$$

and we infer by inspection that the columns (F_{12}, F_{22}) and (G_{11}, G_{21}) are analytic in the upper half λ plane, whereas the columns (F_{11}, F_{21}) and (G_{12}, G_{22}) are analytic in the lower half plane. Assuming, furthermore, that K and N fall off faster than $\exp.(-\lambda_0|y|)$ for $y\to\infty$ and $y\to-\infty$, respectively, we can extend the analyticity domain to include the co-called Bargmann strip $|\mathrm{Im}\lambda| \leq \lambda_0$ (for more details see the succeeding section). Below we have in Fig.6.20 summarised the analytic properties of the Jost functions $F(x\lambda)$ and $G(x\lambda)$.

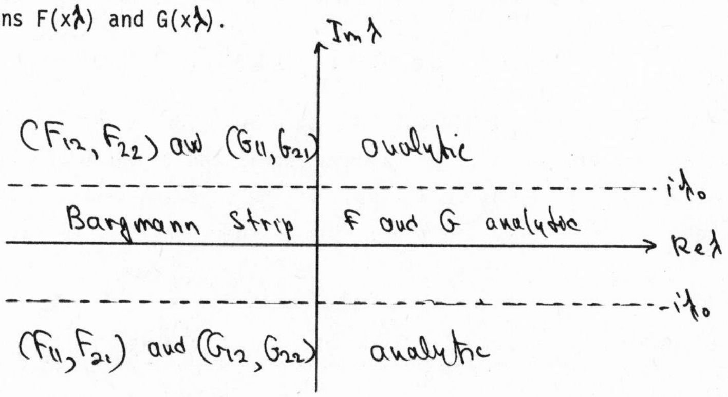

Fig.6.20 Analyticity domain of the Jost functions of F and G

From the general properties (6.35a-c) we conclude that the Jost solutions F and G are related by a constant matrix $T(\lambda)$ only depending on the spectrum $\{\lambda\}$

$$G(x\lambda) = F(x\lambda)T(\lambda). \tag{6.40}$$

The transition matrix $T(\lambda)$ characterises the eigenvalue problem (6.34) and is, using (6.38a-b) formally given by

$$T(\lambda) = \left(\exp\left(-i\lambda\int_{-\infty}^{\infty}(S(x)-\sigma^z)\,dx\right)\right)_+ . \tag{6.41}$$

Since $S(x) \to \sigma^z$ for $|x| \to \infty$ we have detF = detG = 1, i.e., detT = 1 and, moreover, $T(\lambda)^* = \sigma^y T(\lambda^*)\sigma^y$. Denoting the elements of $T(\lambda)$ by $a(\lambda)$ and $b(\lambda)$ the above symmetry properties imply

$$T(\lambda) = \begin{vmatrix} a(\lambda) & -(b(\lambda^*))^* \\ b(\lambda) & (a(\lambda^*))^* \end{vmatrix} , \tag{6.42}$$

where

$$a(\lambda)(a(\lambda^*))^* + b(\lambda)(b(\lambda^*))^* = 1 . \tag{6.43}$$

In terms of the Jost functions the matrix elements $a(\lambda)$ and $b(\lambda)$ are given by

$$a(\lambda) = F_{22}(x\lambda)G_{11}(x\lambda) - F_{12}(x\lambda)G_{21}(x\lambda) \tag{6.44a}$$

$$b(\lambda) = F_{11}(x\lambda)G_{21}(x\lambda) - F_{21}(x\lambda)G_{11}(x\lambda) \tag{6.44b}$$

and we infer from the analytic properties of F and G, as summarised in Fig. 6.20, that $a(\lambda)$ is analytic in the Bargmann strip and in the upper half plane, and that $b(\lambda)$ is analytic at least in the Bargmann strip. In Fig. 6.21, we have shown the analytic properties of the transition matrix $T(\lambda)$

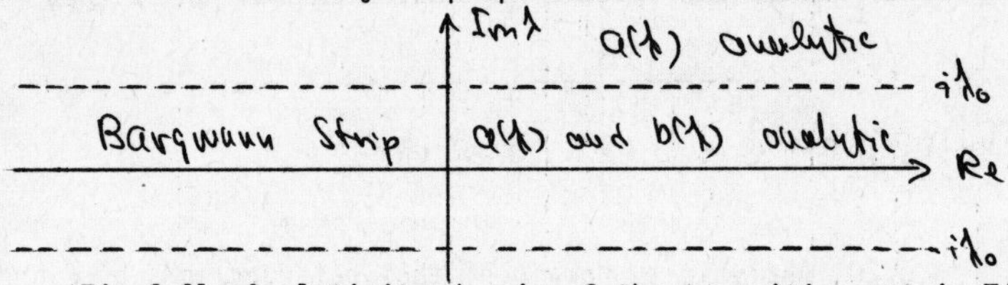

Fig.6.21 Analyticity domain of the transition matrix $T(\lambda)$

6.7 Spectral Properties - Mathematical Discussion

Here we discuss in some detail the spectral properties of the eigenvalue problem, following standard mathematical treatments (see, for instance, L.D. Faddeev, ref. 107).

For $|x| \to \infty$ the general solution of (6.34), $i\frac{d\Psi}{dx} = S\Psi$, has the form $\exp(-i\lambda\sigma^z x)A$, where A is a constant matrix. Using the method of variations of parameters we insert $\Psi(x) = \exp(-i\lambda\sigma^z x)A(x)$ in Eq.(6.34) and obtain for A(x)

$$i\frac{dA(x)}{dx} = \exp(i\lambda\sigma^z x)(S(x) - \sigma^z)\exp(-i\lambda\sigma^z x)A(x)$$

with the solution

$$A(x) = A(x_0) - i\lambda\int_{x_0}^{x}\exp(i\lambda\sigma^z y)(S(y)-\sigma^z)\exp(-i\lambda\sigma^z y)A(y)\,dy,$$

and it follows that $\Psi(x)$ satisfies the Volterra integral equation

$$\Psi(x) = \exp(-i\lambda\sigma^z(x-x_0))\Psi(x_0) - i\lambda\int_{x_0}^{x}\exp(-i\lambda\sigma^z(x-y))(S(y)-\sigma^z)\Psi(y)\,dy$$

For the Jost functions $F(x\lambda)$ and $G(x\lambda)$, in particular, we have

$$F(x\lambda) = \exp(-i\lambda\sigma^z x) + i\lambda\int_{x}^{\infty}\exp(-i\lambda\sigma^z(x-y))(S(y)-\sigma^z)F(y\lambda)\,dy \qquad (6.45a)$$

$$G(x\lambda) = \exp(-i\lambda\sigma^z x) - i\lambda\int_{-\infty}^{x}\exp(-i\lambda\sigma^z(x-y))(S(y)-\sigma^z)G(y\lambda)\,dy\,. \qquad (6.45b)$$

In order to examine the analytic properties of F and G as functions of λ we perform a Neumann expansion[17] of the Volterra integral equations (6.45a - b),

$$F(x\lambda) = \sum_{n=0}^{\infty}F^{(n)}(x\lambda),\quad F^{(0)}(x\lambda) = \exp(-i\lambda\sigma^z x)$$

$$G(x\lambda) = \sum_{n=0}^{\infty}G^{(n)}(x\lambda),\quad G^{(0)}(x\lambda) = \exp(-i\lambda\sigma^z x),$$

where

$$F^{(n+1)}(x\lambda) = i\lambda\int_{x}^{\infty}\exp(-i\lambda\sigma^z(x-y))(S(y)-\sigma^z)F^{(n)}(y\lambda)\,dy$$

$$G^{(n+1)}(x\lambda) = -i\lambda\int_{-\infty}^{x}\exp(-i\lambda\sigma^z(x-y))(S(y)-\sigma^z)G^{(n)}(y\lambda)\,dy$$

It is easy to establish the bounds

$$|F^{(n+1)}(x\lambda)| < |\lambda|\int_{x}^{\infty}\exp(\lambda''\sigma^z(x-y))|S(y)-\sigma^z||F^{(n)}(y\lambda)|\,dy$$

$$|G^{(n+1)}(x\lambda)| < |\lambda|\int_{-\infty}^{x}\exp(\lambda''\sigma^z(x-y))|S(y)-\sigma^z||G^{(n)}(y\lambda)|\,dy$$

Here $|A|$ denotes the matrix $|A_{ij}|$ and λ'' the imaginary part of λ, $\lambda = \lambda' + i\lambda''$. A first iteration yields

$$|F^{(1)}(x,\lambda)| < |\lambda| \exp(\lambda'' \sigma^z x) \int_x^{\infty} V(y)\, dy$$

$$|G^{(1)}(x,\lambda)| < |\lambda| \exp(\lambda'' \sigma^z x) \int_{-\infty}^{x} V(y)\, dy \ ,$$

where

$$V(y) = \begin{vmatrix} 1 - S^z(y) & \exp(-2\lambda'' y)\sqrt{1 - S^z(y)^2} \\ \exp(2\lambda'' y)\sqrt{1 - S^z(y)^2} & 1 - S^z(y) \end{vmatrix}$$

The bounds are onviously controlled by the integrals $\int_x^{\infty} V(y)dy$ and $\int_{-\infty}^{x} V(y)dy$. Provided $S^z(y)$ approaches the ground state value 1 fast enough for $|y| \to \infty$ there is in general a strip in the complex λ plane where the integrals exist. More precisely, if $|1-S^z(y)| < \exp(-4\lambda_0 |y|)$ for $|y| \to \infty$ the integrals are convergent in the Bargmann strip $|\lambda''| < \lambda_0$, see Fig.6.21 in Section 6.6. We notice that under the weaker condition $|1-S^z(y)| < 1/|y|^2$ for $|y| \to \infty$ the second column of $\int_x^{\infty} V(y)dy$ is finite for $\lambda'' > 0$ and the first column finite for $\lambda'' < 0$, and vice versa for $\int_{-\infty}^{x} V(y)dy$. Denoting $\int_x^{\infty} V(y)dy = M(x)$, $M'(x) = -V(x)$, we have

$$|F^{(1)}(x,\lambda)| < |\lambda| \exp(\lambda'' \sigma^z x) M(x)$$

$$|F^{(2)}(x,\lambda)| < |\lambda|^2 \exp(\lambda'' \sigma^z x) \int_x^{\infty} V(y) M(y)\, dy$$

$$= |\lambda|^2 \exp(\lambda'' \sigma^z x) \frac{M(x)^2}{2}$$

and by induction

$$|F^{(n)}(x,\lambda)| < |\lambda|^n \exp(\lambda'' \sigma^z x) \frac{M(x)^n}{n!}$$

Similarly,

$$|G^{(n)}(x,\lambda)| < |\lambda|^n \exp(\lambda'' \sigma^z x) \frac{N(x)^n}{n!} \ ,$$

where $\int_{-\infty}^{x} V(y)dy = N(x)$. From the bounds for $F^{(n)}$ and $G^{(n)}$ we infer

$$F(x\lambda) = \sum_{n=0}^{\infty} F^{(n)}(x\lambda) < \exp(\lambda'' \sigma^3 x) \exp(|\lambda| M(x))$$

$$G(x\lambda) = \sum_{n=0}^{\infty} G^{(n)}(x\lambda) < \exp(\lambda'' \sigma^3 x) \exp(|\lambda| N(x))$$

and it follows that the Volterra integral equations (6.45a-b) have solutions and that, furthermore, because of uniform convergence, the Jost functions F and G are analytic in the Bargmann strip $|\lambda''| < \lambda_0$.

The analyticity domain is actually larger depending on which matrix elements of F and G we consider. Since

$$M^n(x) = n! \int_x^{\infty} dx_1 V(x_1) \int_{x_1}^{\infty} dx_2 V(x_2) \cdots \int_{x_{n-1}}^{\infty} dx_n V(x_n)$$

$$N^n(x) = n! \int_{-\infty}^{x} dx_1 V(x_1) \int_{-\infty}^{x_1} dx_2 V(x_2) \cdots \int_{-\infty}^{x_{n-1}} dx_n V(x_n)$$

we infer by inspection, inserting V(x), that the columns (M^n_{11}, M^n_{21}) and (N^n_{12}, N^n_{22}) are analytic in the lower half plane $\lambda'' < 0$ and the columns (M^n_{12}, M^n_{22}) and (N^n_{11}, N^n_{21}) analytic in the upper half plane $\lambda'' > 0$, under the weaker condition $|1-S^z(y)| < 1/|y|^2$ for $|y| \to \infty$. Consequently, (F_{11}, F_{21}) and (G_{12}, G_{22}) are analytic for $\lambda'' < 0$, and (F_{12}, F_{22}) and (G_{11}, G_{21}) analytic for $\lambda'' > 0$, and we obtain the analyticity domain for the Jost functions shown in Fig.6.20 in Section 6.6.

Having derived the domains of analyticity for the Jost functions F and G we are now in position to deduce the Jost representations (6.39a-b). By inspection of the bounds and by induction it is easy to establish that

$$(F_{11}(x\lambda), F_{21}(x\lambda)) < \exp(-|\lambda''|x)\, O(1) \quad \text{for} \quad \lambda'' < 0$$

$$(G_{12}(x\lambda), G_{22}(x\lambda)) < \exp(-|\lambda''|x)\, O(1) \quad \text{for} \quad \lambda'' < 0$$

$$(F_{12}(x\lambda), F_{22}(x\lambda)) < \exp(-\lambda'' x)\, O(1) \quad \text{for} \quad \lambda'' > 0$$

$$(G_{11}(x\lambda), G_{21}(x\lambda)) < \exp(-\lambda'' x)\, O(1) \quad \text{for} \quad \lambda'' > 0$$

Consequently, by closing the contour in the lower half plane

$$\int \begin{pmatrix} F_{11}(x\lambda) - \exp(-i\lambda x) \\ F_{21}(x\lambda) \end{pmatrix} \frac{\exp(i\lambda y)}{\lambda} \frac{d\lambda}{2\pi} = 0 \quad \text{for} \quad x > y$$

$$\int \begin{pmatrix} F_{12}(x) \\ F_{22}(x\lambda) - \exp(i\lambda x) \end{pmatrix} \frac{\exp(-i\lambda y)}{\lambda} \frac{d\lambda}{2\pi} = 0 \quad \text{for} \quad x > y$$

and similar expressions for $G(x\lambda)$, and we infer in matrix form the Jost representations (6.39a-b),

$$F(x\lambda) = \exp(-i\lambda\sigma^2 x) + \lambda \int_x^{\infty} K(xy) \exp(-i\lambda\sigma^2 y)\, dy$$

$$G(x\lambda) = \exp(-i\lambda\sigma^2 x) + \lambda \int_{-\infty}^{x} N(xy) \exp(-i\lambda\sigma^2 y)\, dy$$

Since both F and G are solutions of the eigenvalue equation (6.34), we obtain, by insertion of the Jost representations, partial differential equations for the kernels K and N,

$$\frac{dK(xy)}{dx}\sigma^2 + S(x)\frac{dK(xy)}{dy} = 0 \quad \text{for } x \leq y \tag{6.46a}$$

$$\frac{dN(xy)}{dx}\sigma^2 + S(x)\frac{dN(xy)}{dy} = 0 \quad \text{for } x \geq y \tag{6.46b}$$

with the boundary conditions

$$S(x) - \sigma^2 + i K(xx) - i S(x) K(xx) \sigma^2 = 0 \tag{6.47a}$$

$$S(x) - \sigma^2 - i N(xx) + i S(x) N(xx) \sigma^2 = 0 , \tag{6.47b}$$

i.e.,

$$S(x) = (iK(xx) - \sigma^2)\, \sigma^2 (iK(xx) - \sigma^2)^{-1} \tag{6.48a}$$

$$S(x) = (i N(xx) + \sigma^2)\, \sigma^2 (i N(xx) + \sigma^2)^{-1} . \tag{6.48b}$$

The initial value problems (6.46a-b and (6.47a-b) for the kernels K and N have been considered by Goursat[108]. The important point in the present context is, of course, that the kernels are independent of the eigenvalue λ.

6.8 Scattering States and Bound States

In analogy with the spectral theory of the one dimensional Schrödinger equation[8,107]

$$-\frac{d^2\psi}{dx^2} + V\psi = E\psi,$$

we can distinguish two kinds of solutions to the eigenvalue problem (6.34)

$$i\frac{d\psi}{dx} = \lambda S\psi$$

Scattering solutions corresponding to a band of real eigenvalues $-\infty<\lambda<\infty$, $\lambda''=0$, and bound state solutions characterised by discrete complex eigenvalues λ_n, n=1,2,...M. The spin field S(x) plays the role of a "potential" giving rise to "wave functions" $\psi(x)$ with different asymptotic character.

The scattering solutions for real λ are characterised by the transmission and reflection of an incoming wave[8]. By means of the Jost solution $G(x\lambda)$ with the boundary condition (6.37b), i.e. $G(x\lambda)\to\exp(-i\lambda\sigma^z x)$ for $x\to-\infty$, we can express the transmitted wave at $x\to-\infty$ in the form

$$\psi_{out}(x) = G(x\lambda)\begin{vmatrix}\frac{1}{a(\lambda)}\\ 0\end{vmatrix} \simeq \begin{vmatrix}\frac{\exp(-i\lambda x)}{a(\lambda)}\\ 0\end{vmatrix} \quad \text{for } x\to-\infty$$

Introducing the transition matrix $T(\lambda)$ given by Eq. (6.42) in order to relate the two Jost functions F and G, i.e., by Eq. (6.40) $G(x\lambda)=F(x\lambda)T(\lambda)$, the incoming and reflected waves are

$$\psi_{in}(x) = F(x\lambda)T(\lambda)\begin{vmatrix}\frac{1}{a(\lambda)}\\ 0\end{vmatrix} \simeq \begin{vmatrix}\exp(-i\lambda x)\\ \frac{b(\lambda)}{a(\lambda)}\exp(i\lambda x)\end{vmatrix} \text{for } x\to\infty$$

and we infer the transmission coefficient

$$t(\lambda) = \frac{1}{a(\lambda)} \tag{6.49}$$

and the reflection coefficient

$$r(\lambda) = \frac{b(\lambda)}{a(\lambda)} \tag{6.50}$$

in terms of the matrix elements of $T(\lambda)$. The determinantal condition (6.43), $\det T(\lambda) = 1$, furthermore, implies the relationship

$$|t(\lambda)|^2 - |r(\lambda)|^2 = 1 . \tag{6.51}$$

We notice that unlike the Schrödinger equation for a quantum mechanical wave function, the eigenvalue problem considered here is not Hermitian and there is as a result no conservation of probability, as indicated by Eq. (6.51). In Fig. 6.22 we have shown the scattering solution.

The bound state solutions correspond to complex values of λ. By means of the Jost function $G(x\lambda)$, whose first column (G_{11}, G_{21}) is analytic in the upper half plane, we can express a bound state solution, decaying as $\exp(-i\lambda x)$ for $x \to -\infty$ and $\lambda'' > 0$, in the form

$$\psi_{bs}(x) = G(x\lambda)\begin{vmatrix} 1 \\ 0 \end{vmatrix} \simeq \begin{vmatrix} \exp(-i\lambda x) \\ 0 \end{vmatrix} \quad \text{for } x \to -\infty$$

For $x \to \infty$ we obtain, using Eq. (6.40), i.e., $G = FT$,

$$\psi_{bs}(x) = F(x\lambda)\,T(\lambda)\begin{vmatrix} 1 \\ 0 \end{vmatrix} \simeq \begin{vmatrix} a(\lambda)\exp(-i\lambda x) \\ b(\lambda)\exp(i\lambda x) \end{vmatrix} \text{for } x \to \infty$$

In order to exclude an exponentially increasing solution for $x \to \infty$ we require $a(\lambda) = 0$, that is the bound state spectrum is determined by the zeroes of $a(\lambda)$ in the upper half complex λ plane. Since $a(\lambda)$ is analytic for $\lambda'' > 0$ and moreover, according to the Jost representations (6.39a-b) and Eq.(6.44a), $a(\lambda) \to 1$ for $|\lambda| \to \infty$, we conclude that $a(\lambda)$ has only a finite number of zeroes λ_n, n = 1,2,..M. The bound state solutions are furthermore specified by the "asymptotic characteristics" b_n,

$$\psi_{bs}(x) \simeq b_n \exp(i\lambda_n x) \text{ for } x \to \infty , n=1,2,\dots M .$$

In case λ_n falls within the Bargmann strip we have, of course, by analytic continuation

$$b_n = b(\lambda_n) \quad \text{for} \quad -\lambda_0 < \lambda_n'' < \lambda_0 \quad , n=1,2,\dots M.$$

In Fig. 6.23 we have sketched the bound state solution

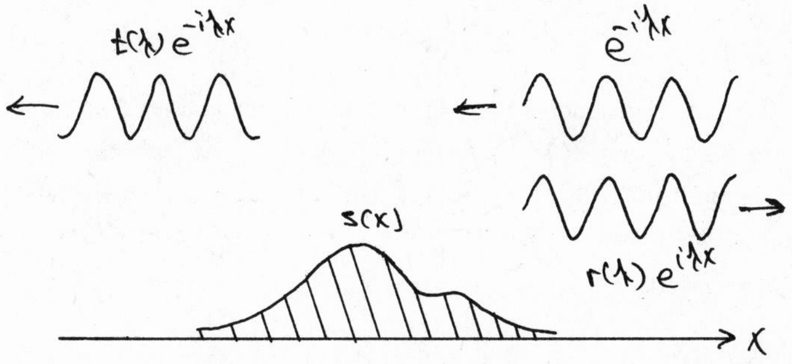

Fig. 6.22 Transmitted and reflected waves for real λ, characterising a scattering solution

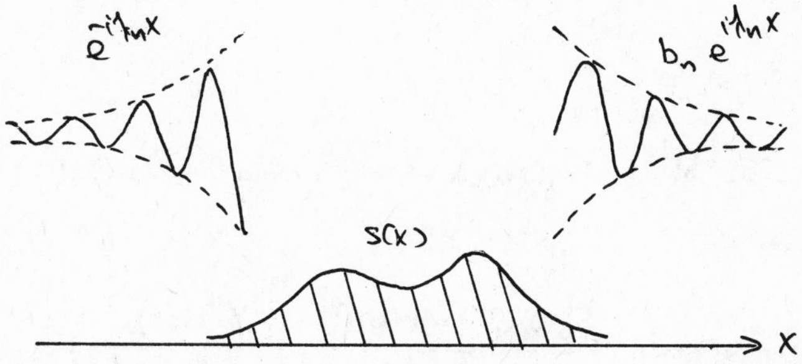

Fig. 6.23 Bound state solution for complex λ

The scattering and bound state solutions of the eigenvalue equation (6.34) are completely characterised by the scattering data,

$$\{r(\lambda), -\infty<\lambda<\infty ; \lambda_n, b_n, n=1,2,\cdots M\}, \qquad (6.52)$$

which in turn uniquely determines the transition matrix $T(\lambda)$. For later purposes we establish this relationship.

Consider the function

$$\bar{a}(\lambda) = a(\lambda)\prod_{n=1}^{M}\left(\frac{\lambda-\lambda_n^*}{\lambda-\lambda_n}\right),$$

which is analytic in the upper half plane, has no zeroes, and approaches unity for $|\lambda|\to\infty$. Since $\bar{a}(\lambda)$ is an entire function it follows that $\log\bar{a}(\lambda)$ is analytic for $\lambda''>0$ and approaches zero for $|\lambda|\to\infty$. On the real axis $\log\bar{a}(\lambda)$ thus satisfies the Kramers-Kronig relations (see, for instance, ref. 21)

$$(\log\bar{a}(\lambda))' = P\int\frac{d\mu}{\pi}\frac{(\log\bar{a}(\mu))''}{\mu-\lambda},$$

$$(\log\bar{a}(\lambda))'' = -P\int\frac{d\mu}{\pi}\frac{(\log\bar{a}(\mu))'}{\mu-\lambda}. \qquad \text{(P principal value)}$$

On the real axis, using Eq. (6.50),

$$\log|\bar{a}(\lambda)| = \log|a(\lambda)| = -\frac{1}{2}\log(1+|r(\lambda)|^2),$$

and we obtain

$$a(\lambda) = \exp\left(-\int\frac{d\mu}{2\pi i}\frac{\log(1+|r(\mu)|^2)}{\mu-\lambda-i\varepsilon}\right)\prod_{n=1}^{M}\left(\frac{\lambda-\lambda_n}{\lambda-\lambda_n^*}\right), \qquad (6.53)$$

which together with Eq. (6.50),

$$b(\lambda) = r(\lambda)a(\lambda),$$

express the matrix elements $a(\lambda)$ and $b(\lambda)$, that is the transition matrix $T(\lambda)$, in terms of the scattering data (6.52). The spectral representation (6.53) shows that $a(\lambda)$ has a branch cut along the real axis corresponding to the scattering states, the discontinuity across the cut being determined by the reflection coefficient $r(\lambda)$, and moreover allows for an analytic continuation onto the second Riemann sheet where $a(\lambda)$ has poles at $\lambda = \lambda^*_n$, n=1,2,..M. In Fig. 6.24 we have summarised the analytic properties of $a(\lambda)$

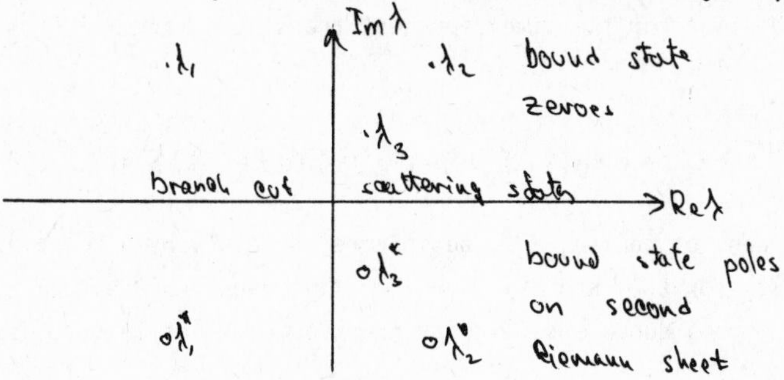

Fig. 6.24 The analytic properties of $a(\lambda)$

6.9 Time Dependence of the Scattering Data

In order to determine the time dependence of the scattering data (6.52) or, equivalently, the transition matrix $T(\lambda)$, induced by the time development of the "potential" $S(xt)$ due to the non linear evolution equation (6.31), we consider a solution at $x \to -\infty$ of the form

$$\psi(xt) = G(x\lambda)A(t) \quad \text{for } x \to -\infty$$

By means of the equation of motion (6.33),

$$i\frac{d\psi(xt)}{dt} = 2S(xt)\frac{d^2\psi(xt)}{dx^2} + \frac{dS(xt)}{dx}\frac{d\psi(xt)}{dx}$$

and the boundary conditions $S(xt) \to \sigma^z$, $dS(xt)/dx \to 0$ for $|x| \to \infty$ and $G(x\lambda) \to \exp(-i\lambda\sigma^z x)$ for $x \to -\infty$, we have

$$A(t) = \exp(2i\lambda^2\sigma^z t)A(0)$$

In the asymptotic region $x \to \infty$, introducing the transition matrix $T(\lambda)$ and using $G(x\lambda) = F(x\lambda)T(\lambda)$, the wave function $\psi(xt)$ behaves as

$$\psi(xt) \sim F(x\lambda)\,T(\lambda t)\exp(2i\lambda^2\sigma^z t)A(0).$$

Again applying the equation of motion (6.32) together with the boundary condition $F(x\lambda) \to \exp(-i\lambda\sigma^z x)$ for $x \to \infty$ we arrive at a first order differential equation for the time dependent transition matrix $T(\lambda t)$ with the solution

$$T(\lambda t) = \exp(2i\lambda^2\sigma^z t)\,T(\lambda 0)\exp(-2i\lambda^2\sigma^z t). \tag{6.54}$$

This result can, of course, also be inferred directly by inspection of the form of $\psi(xt)$ for $t \to \infty$ and $t \to -\infty$. The time dependence of the "potential" $S(xt)$ thus induces a similarity transformation of $T(\lambda t)$. Using

$$\exp(2i\lambda^2\sigma^z t) = \cos(2\lambda^2 t) + i\sigma^z \sin(2\lambda^2 t)$$

we obtain for the matrix elements $a(\lambda t)$ and $b(\lambda t)$

$$a(\lambda t) = a(\lambda 0) \tag{6.55a}$$

$$b(\lambda t) = b(\lambda 0)\exp(-4i\lambda^2 t) \tag{6.55b}$$

or, equivalently, for the scattering data (6.52)

$$r(\lambda t) = r(\lambda 0)\exp(-4i\lambda^2 t), \quad -\infty < \lambda < \infty \tag{6.56a}$$

$$\lambda_n(t) = \lambda_n(0), \qquad n = 1,2,\ldots M \tag{6.56b}$$

$$b_n(t) = b_n(0)\exp(-4i\lambda_n^2 t), \quad n = 1,2,\ldots M \tag{6.56c}$$

in accordance with ref. 102. We remark that, in conformity with the general discussion in Section 6.5, the bound state spectrum $\{\lambda_n\}$, n = 1,2..M. is independent of time.

6.10 The Inverse Scattering Problem

The reconstruction of the spin field, i.e., the "potential" $S(xt)$, from the time dependent scattering data (6.56a-c) is usually called the "inverse scattering problem", and is achieved by means of a linear integral equation, the Gelfand-Levitan-Marchenko equation (see, for instance, refs. 8 and 1o7), which we now proceed to derive.

From Eq. (6.40), $G(x\lambda) = F(x\lambda)T(\lambda)$, we have, using Eq. (6.42),

$$G_{11}(x\lambda) = F_{11}(x\lambda)\,a(\lambda) + F_{12}(x\lambda)\,b(\lambda)$$

$$G_{21}(x\lambda) = F_{21}(x\lambda)\,a(\lambda) + F_{22}(x\lambda)\,b(\lambda)$$

or, since $a(\lambda) \to 1$ for $|\lambda| \to \infty$ in the upper half plane and introducing the reflection coefficient $r(\lambda) = b(\lambda)/a(\lambda)$,

$$\left(\frac{1}{a(\lambda)} - 1\right) G_{11}(x\lambda) = F_{11}(x\lambda) + F_{12}(x\lambda)\,r(\lambda) - G_{11}(x\lambda)$$

$$\left(\frac{1}{a(\lambda)} - 1\right) G_{21}(x\lambda) = F_{21}(x\lambda) + F_{22}(x\lambda)\,r(\lambda) - G_{21}(x\lambda)$$

Inserting the Jost representations (6.39a-b), multiplying by $\exp(i\lambda z)/\lambda$, and integrating over λ, we obtain

$$\int_{-\infty}^{\infty} G_{11}(x\lambda)\left(\frac{1}{a(\lambda)} - 1\right)\frac{e^{i\lambda z}}{\lambda}\frac{d\lambda}{2\pi} =$$

$$\int_{-\infty}^{\infty} e^{i\lambda z}\left(\int_x^{\infty} K_{11}(xy)e^{-i\lambda y}dy + r(\lambda)\int_x^{\infty} K_{12}(xy)e^{i\lambda y}dy - \int_{-\infty}^{x} N_{11}(xy)e^{-i\lambda y}dy\right)\frac{d\lambda}{2\pi}$$

$$\int_{-\infty}^{\infty} G_{21}(x\lambda)\left(\frac{1}{a(\lambda)} - 1\right)\frac{e^{i\lambda z}}{\lambda}\frac{d\lambda}{2\pi} = \int_{-\infty}^{\infty} e^{i\lambda z}\left(\int_x^{\infty} K_{21}(xy)e^{-i\lambda y}dy + \right.$$

$$\left. r(\lambda)\frac{e^{i\lambda x}}{\lambda} + r(\lambda)\int_x^{\infty} K_{22}(xy)e^{i\lambda y}dy - \int_{-\infty}^{x} N_{21}(xy)e^{-i\lambda y}dy\right)\frac{d\lambda}{2\pi}$$

Since $G_{11}(x\lambda)$ and $G_{21}(x\lambda)$ are analytic in the upper half plane (see Fig. 6.20) and fall off faster than $\exp(-\lambda''x)$ we can for $z > -x$ close the contours for $\lambda'' > 0$. The integrals on the right hand sides are straightforward and we have, using $\int \exp(i\lambda x)\frac{d\lambda}{2\pi} = \delta(x)$,

$$i\sum_{n=1}^{M}\frac{e^{i\lambda_n z}G_{11}(x\lambda_n)}{\lambda_n a'(\lambda_n)} = K_{11}(xz) - N_{11}(xz) + \int_x^\infty K_{12}(xy)\left(\int_{-\infty}^{\infty} e^{i\lambda(y+z)} r(\lambda)\frac{d\lambda}{2\pi}\right)dy$$

$$i\sum_{n=1}^{M}\frac{e^{i\lambda_n z}G_{21}(x\lambda_n)}{\lambda_n a'(\lambda_n)} = K_{21}(xz) - N_{21}(xz) + \int_x^\infty K_{22}(xy)\left(\int_{-\infty}^{\infty} e^{i\lambda(y+z)} r(\lambda)\frac{d\lambda}{2\pi}\right)dy$$

$$+ \int_{-\infty}^{\infty}\frac{d\lambda}{2\pi}\, r(\lambda)\frac{e^{i\lambda(x+z)}}{\lambda},$$

where $a'(\lambda_n) = \left(da(\lambda)/d\lambda\right)_{\lambda=\lambda_n}$. At the bound state zeroes we obtain, by means of $G(x\lambda) = F(x\lambda)T(\lambda)$ and the Jost representations (6.39a-b)

$$G_{11}(x\lambda) = b_n\lambda_n\int_x^\infty K_{12}(xy)\, e^{i\lambda_n y}dy$$

$$G_{21}(x\lambda) = b_n e^{i\lambda_n x} + b_n\lambda_n\int_x^\infty K_{22}(xy)\, e^{i\lambda_n y}dy$$

which by insertion, for $z > x$ and denoting $b_b/a'(\lambda_n) = im_n$, yield the integral equations

$$K_{11}(xz) + \int_x^\infty K_{12}(xy)\left(\int_{-\infty}^{\infty} e^{i\lambda(y+z)} r(\lambda)\frac{d\lambda}{2\pi} + \sum_{n=1}^{M} e^{i\lambda_n(y+z)} m_n\right)dy = 0 \quad (6.57a)$$

$$K_{21}(xz) + \left(\int_{-\infty}^{\infty} e^{i\lambda(x+z)}\frac{r(\lambda)}{\lambda}\frac{d\lambda}{2\pi} + \sum_{n=1}^{M} e^{i\lambda_n(x+z)}\frac{m_n}{\lambda_n}\right) +$$
$$\int_x^\infty K_{22}(xy)\left(\int_{-\infty}^{\infty} e^{i\lambda(y+z)} r(\lambda)\frac{d\lambda}{2\pi} + \sum_{n=1}^{M} e^{i\lambda_n(y+z)} m_n\right)dy = 0. \quad (6.57b)$$

In a similar manner, another set of integral equations can be derived for the elements K_{12} and K_{22}. Compactly, the Gelfand-Levitan-Marchenko equation for the Jost kernel K can be expressed in the form

$$K(xyt) + \Phi_1(x+y\,t) + \int_x^\infty K(xzt)\,\Phi_2(z+y\,t)\,dz = 0 \quad (6.58)$$

for $x \le y$,

where Φ_1 and Φ_2 are functions of the time dependent scattering data,

$$\Phi_1(xt) = i\,\mathrm{Im}F(xt)\sigma^x - i\,\mathrm{Re}F(xt)\sigma^y \tag{6.59a}$$

$$\Phi_2(xt) = i\sigma^z \frac{d\Phi_1(xt)}{dx} \tag{6.59b}$$

$$F(xt) = \int r(\lambda t)\frac{e^{i\lambda x}}{\lambda}\frac{d\lambda}{2\pi} + \sum_{n=1}^{M}\frac{b_n(t)}{i a'(\lambda_n)\lambda_n}\, e^{i\lambda_n x}. \tag{6.60}$$

We have chosen the same notation as in ref. 102, where the Gelfand-Levitan-Marchenko equation is presented but not derived. Having in principle solved equation (6.58) for a given set of scattering data (6.52), the spin density is subsequently determined by Eq.(6.48a),

$$S(xt) = (iK(xxt) - \sigma^z)\,\sigma^z\,(iK(xxt) - \sigma^z)^{-1}.$$

The important point about the Gelfand-Levitan-Marchenko integral equation is not that it is generally soluble, it is not! But the fact that it is linear and therefore lends itself to analysis and approximation schemes. An important exception is, however, the case of a "reflectionsless" potential, i.e., $r(\lambda t) = 0$, where the integral equation reduces to a set of linear algebraic equations. This case corresponds to the multi soliton solutions (see later) which thus can be derived explicitly.

Schematically, we can illustrate the "inverse scattering approach" to non linear evolution equations by means of the diagram[89]

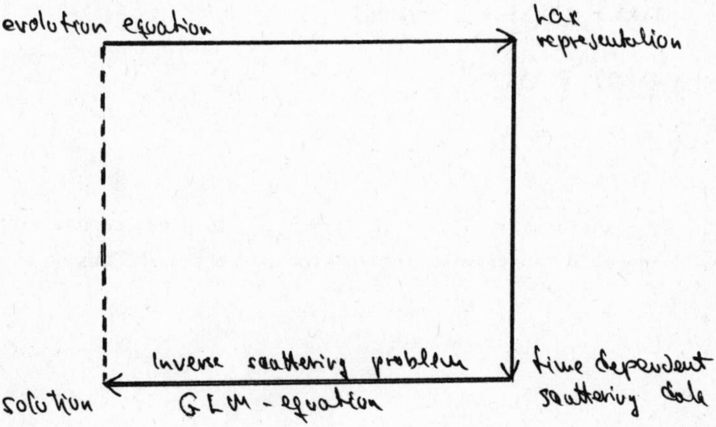

Here the dashed line indicates the direct but in general intractable route to the solution. The weak point in the "inverse scattering method" applied to non linear evolution equations is, of course, that there in general is no systematic way of constructing the Lax representation, which has to be guessed. Notice, however, some recent work by Lamb[109,110] on moving space curves associated with some of the non linear equations, and also by Lakshmanan[111] concerning the spin problem discussed here.

6.11 Canonical Action Angle Variables

In the previous section we derived the Gelfand-Levitan-Marchenko equation (6.58) for the inverse scattering problem associated with the eigenvalue equation (6.34), thus providing an explicit procedure for the construction of solutions to the non linear precessional equation of motion (6.29). In the present context we shall, however, not consider explicit solutions, but employ the "inverse scattering formalism" developed in the preceeding sections in order to construct canonical action angle variables[68] for the classical Heisenberg chain.

In Section 6.2 we gave a canonical representation of the Heisenberg chain. The spin Hamiltonian (6.1) for h = 0 takes the form (6.6)

$$H = \frac{1}{2}\int dx\left(\frac{1}{1-p^2}\left(\frac{dp}{dx}\right)^2 + (1-p^2)\left(\frac{dq}{dx}\right)^2\right),$$

where the canonical variables $p = S^z = \cos\Theta$ and $q = \tan^{-1}(S^y/S^x) = \Phi$ obey the Poisson bracket relations (6.4a-c),

$$\{p(x), q(y)\} = \delta(x-y)$$

$$\{p(x), p(y)\} = 0$$

$$\{q(x), q(y)\} = 0.$$

A canonical transformation[68] $p(x), q(x) \rightarrow P_n, Q_n$ to a new canonical basis has the property of preserving the Poisson bracket relations,

$$\{P_n, Q_m\} = \delta_{nm}$$

$$\{P_n, P_m\} = 0$$

$$\{Q_n, Q_m\} = 0 .$$

Here n is a general multi dimensional functional index. In particular, the equations of motion

$$\frac{dp}{dt} = \{H, p\} = -\frac{dH}{dq}$$

$$\frac{dq}{dt} = \{H, q\} = \frac{dH}{dp}$$

are transformed into (suppressing the index n)

$$\frac{dP}{dt} = \{H', P\} = -\frac{dH'}{dQ}$$

$$\frac{dQ}{dt} = \{H', Q\} = \frac{dH'}{dP} ,$$

where $H'(PQ) = H(pq)$ is the new Hamiltonian.

The inverse scattering method allows for the construction of that particular canonical transformation which yields a Hamiltonian $H'(P)$ only depending on the new canonical momentum P. The Hamiltonian equations of motion are then immediately soluble and imply the action angle representation[68]

$$P = \text{const.}$$

$$Q = \left(\frac{dH'}{dP}\right) t + \text{const.},$$

i.e., the canonical momentum P is a constant of motion, and the canonical coordinate Q develops linearly in time with a characteristic frequency $\omega = dH'/dP$.

For the purpose of constructing canonical action angle variables for the Heisenberg chain the easiest way to proceed is to derive the Poisson bracket relations for the scattering data (6.52),

$$\{r(\lambda), -\infty<\lambda<\infty;\ \lambda_n, b_n, n=1,2,\dots M\},$$

or, equivalently, for the transition matrix $T(\lambda)$. We here follow Zakharov and Manakov[112] who have applied such techniques to the non linear Schrödinger equation and to the Korteweg-deVries equation.

By definition[68] (suppressing the time variable)

$$\{T_{ij}(\lambda), T_{mn}(\mu)\} = \int dx\left(\frac{dT_{ij}(\lambda)}{dp(x)}\frac{dT_{mn}(\mu)}{dq(x)} - \frac{dT_{ij}(\lambda)}{dq(x)}\frac{dT_{mn}(\mu)}{dp(x)}\right)$$

or, transforming to the equivalent spin variables $S^\alpha(x)$, and using the Poisson bracket relation (6.2), $\{S^\alpha(x), S^\beta(y)\} = -\delta(x-y)\sum_\gamma \varepsilon^{\alpha\beta\gamma}S^\gamma(x)$,

$$\{T_{ij}(\lambda), T_{mn}(\mu)\} = -\sum_{\alpha\beta\gamma}\int dx\,\frac{dT_{ij}(\lambda)}{dS^\alpha(x)}\frac{dT_{mn}(\mu)}{dS^\beta(x)}\varepsilon^{\alpha\beta\gamma}S^\gamma(x).$$

In order to evaluate the derivative $dT_{ij}(\lambda)/dS^\alpha(x)$ we make use of the definition (6.40), $T(\lambda) = F^{-1}(y\lambda)G(y\lambda)$, y arbitrary, together with the formal identity $\delta F^{-1}F + F^{-1}\delta F = 0$, i.e.,

$$\frac{dT(\lambda)}{dS^\alpha(x)} = F^{-1}(y\lambda)\frac{dG(y\lambda)}{dS^\alpha(x)} - F^{-1}(y\lambda)\frac{dF(y\lambda)}{dS^\alpha(x)}T(\lambda).$$

From the Jost representations (6.39a-b) we conclude, however, that $dF(y\lambda)/dS^\alpha(x) = dG(x\lambda)/dS^\alpha(y) = 0$ for $y > x$. Choosing $y > x$ we thus obtain $dT(\lambda)/S^\alpha(x) = F^{-1}(y\lambda)dG(y\lambda)/dS^\alpha(x)$ for $y > x$. Since the Jost function $G(y\lambda)$ is a solution of the eigenvalue equation (6.34), $idG(y\lambda)/dy = \lambda S(y)G(y\lambda)$, we infer by differentiation with respect to $S^\alpha(x)$ that $dG(y\lambda)/dS^\alpha(x)$ is determined by the Green's function equation

$$\left(i\frac{d}{dy} - \lambda S(y)\right)\frac{dG(y\lambda)}{dS^\alpha(x)} = \lambda\sigma^\alpha G(y\lambda)\delta(x-y).$$

Following the standard procedure[17], i.e., integrating over the inverval $x-\varepsilon$ to $x+\varepsilon$ and letting $\varepsilon \to 0$, we obtain the boundary condition $idG(x^+\lambda)/dS^\alpha(x) = \lambda\sigma^\alpha G(x\lambda)$ and in terms of $G(y\lambda)$ the solution

$$\frac{dG(y\lambda)}{dS^\alpha(x)} = -i\lambda G(y\lambda)G^{-1}(x\lambda)\sigma^\alpha G(x\lambda) \quad \text{for } y > x$$

Inserting dG/dS^{α} in the expression for dT/dS^{α} and choosing $y = x^{+}$ we have

$$\frac{dT(\lambda)}{dS^{\alpha}(x)} = -i\lambda F^{-1}(x\lambda)\sigma^{\alpha}G(x\lambda). \tag{6.61}$$

This result can also be obtained by derivation of the formal expression (6.41) with respect to $S^{\alpha}(x)$, using (6.38a-b). Substituting Eq.(6.61) in the expression for $\{T(\lambda),T(\mu)\}$ and using the identity

$$\frac{d}{dx}\Big(\sum_{\alpha}(F^{-1}(x\lambda)\sigma^{\alpha}G(x\lambda))_{ij}(F^{-1}(x\mu)\sigma^{\alpha}G(x\mu))_{mn}\Big) =$$

$$2(\lambda-\mu)\sum_{\alpha\beta\gamma}\varepsilon^{\alpha\beta\gamma}(F^{-1}(x\lambda)\sigma^{\alpha}G(x\lambda))_{ij}(F^{-1}(x\mu)\sigma^{\beta}G(x\mu))_{mn}S^{\gamma}(x),$$

we obtain

$$(\lambda-\mu)\{T_{ij}(\lambda),T_{mn}(\mu)\} =$$

$$\frac{1}{2}\lambda\mu\int\frac{d}{dx}\Big(\sum_{\alpha}(F^{-1}(x\lambda)\sigma^{\alpha}G(x\lambda))_{ij}(F^{-1}(x\mu)\sigma^{\alpha}G(x\mu))_{mn}\Big)dx.$$

Since the integrand is a total derivative we have

$$(\lambda-\mu)\{T_{ij}(\lambda),T_{mn}(\mu)\} = \frac{1}{2}\lambda\mu\lim_{L\to\infty}\sum_{\alpha}\Big((F^{-1}(L\lambda)\sigma^{\alpha}G(L\lambda))_{ij}(F^{-1}(L\mu)\sigma^{\alpha}G(L\mu))_{mn}$$

$$-(F^{-1}(-L\lambda)\sigma^{\alpha}G(-L\lambda))_{ij}(F^{-1}(-L\mu)\sigma^{\alpha}G(-L\mu))_{mn}\Big),$$

where we have controlled the limits of integration by means of a cut off L. From the Jost representations (6.39a-b), the definition $G(x\lambda) = F(x\lambda)T(\lambda)$, and for λ on the real axis

$$F^{-1}(L\lambda) \simeq \exp(i\lambda\sigma^{z}L) \qquad \text{for } L\to\infty$$
$$F^{-1}(-L\lambda) \simeq T(\lambda)\exp(-i\lambda\sigma^{z}L) \qquad \text{for } L\to\infty$$
$$G(L\lambda) \simeq \exp(-i\lambda\sigma^{z}L)\,T(\lambda) \qquad \text{for } L\to\infty$$
$$G(-L\lambda) \simeq \exp(i\lambda\sigma^{z}L), \qquad \text{for } L\to\infty$$

which by insertion in the above expression for $\{T(\lambda),T(\mu)\}$, using

$$\lim_{L\to\infty} P\,\frac{\exp(i\lambda L)}{\lambda} = \pi\delta(\lambda),$$

introducing $\sigma^{\pm} = \sigma^x \pm i\sigma^y$ and performing a bit of Pauli matrix algebra, leads to

$$-\{T_{ij}(\lambda), T_{mn}(\mu)\} = \frac{1}{2} P \frac{\lambda\mu}{\lambda-\mu}\left((\sigma^z T(\lambda))_{ij}(\sigma^z T(\mu))_{mn} - (T(\lambda)\sigma^z)_{ij}(T(\mu)\sigma^z)_{mn}\right)$$

$$-+ i\pi\lambda^2\delta(\lambda-\mu)\Big(((\sigma^+ T(\lambda))_{ij}(\sigma^- T(\mu))_{mn} - (\sigma^- T(\lambda))_{ij}(\sigma^+ T(\mu))_{mn}$$

$$-+ (T(\lambda)\sigma^+)_{ij}(T(\mu)\sigma^-)_{mn} - (T(\lambda)\sigma^-)_{ij}(T(\mu)\sigma^+)_{mn}\Big)$$

This is an important intermediate result which shows that the transition matrix elements $T_{ij}(\lambda)$ satisfy a closed Poisson bracket algebra. All direct reference to configuration space has disappeared and the dynamical behaviour of the Heisenberg chain is now reflected in the Poisson bracket relations (6.62).

For the matrix elements $a(\lambda)$ and $b(\lambda)$ characterising the scattering states, i.e., for λ on the real axis, we infer by inspection of (6.62) the non vanishing Poisson brackets

$$\{a(\lambda), b(\mu)\} = -\lambda\mu a(\lambda) b(\mu)\left(P\frac{1}{\lambda-\mu} - i\pi\delta(\lambda-\mu)\right)$$

$$\{a(\lambda), b(\mu)^*\} = \lambda\mu a(\lambda) b(\mu)^*\left(P\frac{1}{\lambda-\mu} - i\pi\delta(\lambda-\mu)\right)$$

$$\{b(\lambda), b(\mu)^*\} = 2\lambda^2 |a(\lambda)|^2 i\pi\delta(\lambda-\mu).$$

We notice that the two first Poisson brackets can be written in the form

$$\{a(\lambda), b(\mu)\} = -\lambda\mu a(\lambda) b(\mu) \frac{1}{\lambda-\mu+i\varepsilon}$$

$$\{a(\lambda), b(\mu)^*\} = \lambda\mu a(\lambda) b(\mu)^* \frac{1}{\lambda-\mu+i\varepsilon}$$

which, in accordance with the spectral properties of $T(\lambda)$, again shows that $a(\lambda)$ can be analytically continued in the upper half complex λ plane.

In order to derive the Poisson bracket relations for the scattering data pertaining to the bound state spectrum, we make use of the implicit equation

$$a(\lambda_n, \{S\}) = 0, \quad n = 1,2,..M,$$

where we have indicated the functional dependence on the "potential" S(xt). By implicit differentiation with respect to S we obtain

$$\left(\frac{da}{d\lambda}\right)_{\lambda_n}\left(\frac{d\lambda_n}{dS}\right) + \left(\frac{da}{dS}\right)_{\lambda_n} = 0,$$

i.e.

$$\frac{d\lambda_n}{dS} = -\frac{1}{\left(\frac{da}{d\lambda_n}\right)_{\lambda_n}}\left(\frac{da}{dS}\right)_{\lambda_n}.$$

Using the expression (6.62) for λ and μ in the Bargmann strip (see Section 6.6) and the analytic continuation of the above Poisson brakckets for $a(\lambda)$ and $b(\mu)$, it is easy to show that the only non vanishing Poisson bracket of λ_n and $b_n = b(\lambda_n)$ is

$$\{\lambda_n, b_n\} = -\frac{1}{\left(\frac{da}{d\lambda}\right)_{\lambda_n}}\{a(\lambda_n), b_n\} = -\frac{1}{\left(\frac{da}{d\lambda}\right)_{\lambda_n}}\lim_{\lambda\to\lambda_n}\{a(\lambda), b_n\} = \lambda_n^2 b_n.$$

We have here, in order to define the analytic continuation of $b(\lambda)$, chosen the bound state zeroes λ_n within the Bargmann strip.

By means of the above Poisson bracket relations for the elements of the transition matrix $T(\lambda)$, we are now in position to construct a new canonical basis for the Heisenberg chain. The variables associated with the scattering states for λ on the real axis are given by

$$P(\lambda) = -\frac{1}{\pi\lambda^2}\log|a(\lambda)|, \quad P(\lambda) \geq 0, \quad -\infty < \lambda < \infty \tag{6.63a}$$

$$Q(\lambda) = -\operatorname{Arg} b(\lambda), \quad -2\pi \leq Q(\lambda) \leq 0, \quad -\infty < \lambda < \infty \tag{6-63b}$$

and satisfy the canonical Poisson brackets

$$\{P(\lambda), Q(\mu)\} = \delta(\lambda-\mu) \tag{6.64a}$$

$$\{P(\lambda), P(\mu)\} = 0 \tag{6.64b}$$

$$\{Q(\lambda), Q(\mu)\} = 0 \tag{6.64c}$$

The real canonical momentum $P(\lambda)$, $-\infty<\lambda<\infty$, depends only on $a(\lambda)$ and is therefore, according to Eq. (6.55a), a constant of motion. The constraint $|a(\lambda)|^2 = 1-|b(\lambda)|^2 \leq 1$ given by Eq. (6.43) furthermore implies that $P(\lambda)$ has a positive range. The real canonical coordinate $Q(\lambda)$, $-\infty<\lambda<\infty$, is defined as the negative phase of $b(\lambda)$ and is therefore an angle specified to within a multiplum of 2π. From the time dependence of $b(\lambda t)$ given by Eq. (6.54b) we obtain

$$Q(\lambda t) - Q(\lambda 0) = 4\lambda^2 t, \quad -\infty<\lambda<\infty \tag{6.65}$$

i.e., $Q(\lambda t)$ evolves linearly with time with the frequency $\omega(\lambda) = 4\lambda^2$. In Fig. 6.25 we have shown the time dependence of the canonical coordinate $Q(\lambda t)$

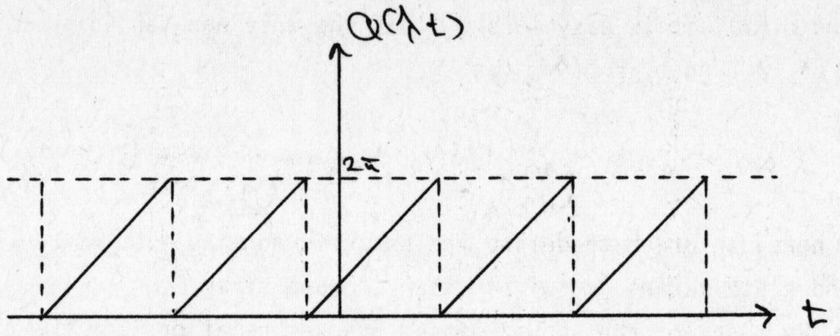

Fig. 6.25 The behaviour of the time dependent phase $Q(\lambda t)$, modulus 2π.

The canonical variables $P(\lambda)$ and $Q(\lambda t)$ are thus of the usual action angle type[68], characteristic of a constrained system.

The variables pertaining to the bound state spectrum for λ in the upper half complex plane are defined by

$$P_n = \frac{i}{\lambda_n}, \quad \mathrm{Re}\, P_n > 0, \quad n=1,2,\cdots M. \tag{6.66a}$$

$$Q_n = i \log b_n, \quad n=1,2,\cdots M. \tag{6.66b}$$

and obey the canonical Poisson brackets

$$\{P_n, Q_m\} = \delta_{nm} \tag{6.67a}$$

$$\{P_n, P_m\} = 0 \tag{6.67b}$$

$$\{Q_n, Q_m\} = 0 . \tag{6.67c}$$

The complex canonical momentum P_n, n = 1,2,..M, is a constant of motion. Since, excluding zeroes on the real axis, Im $\lambda_n > 0$, the range of ReP_n is restricted to positive values. The time dependence of the complex canonical coordinate Q_n, n =1,2..M, is derived from Eq. (6.56c), i.e.,

$$Q_n(t) - Q_n(0) = 4\lambda_n^2 t = -\frac{4}{P_n^2} t , \quad n=1,2,..M$$

The discrete complex canonical variables P_n and Q_n are therefore also of the action angle type[68].

We finally demonstrate that the transition matrix T(λ) is uniquely determined by the canonical action angle variables P(λ), Q(λ), P_n and Q_n. By means of the spectral representation (6.53) together with Eq. (6.50), i.e., r(λ) = b(λ)/a(λ), and the above definitions of the canonical variables, we obtain

$$a(\lambda) = \exp\left(i\int d\mu \frac{\mu^2 P(\mu)}{\mu-\lambda-i\varepsilon}\right)\prod_{n=1}^{M}\left(\frac{\lambda - iP_n^{-1}}{\lambda + iP_n^{-1*}}\right) \tag{6.69a}$$

$$b(\lambda) = (1-\exp(-2\pi\lambda^2 P(\lambda)))^{\frac{1}{2}} \exp(-iQ(\lambda)) \tag{6.69b}$$

$$b_n = \exp(-iQ_n) \tag{6.69c}$$

In the above paragraphs we have derived the canonical action angle variables for the classical Heisenberg chain in the long wavelength limit. The inverse scattering method developed in the previous sections essentially allows for the implicit construction of the non linear canonical transformation S(xt)$\rightarrow$ P(λ),Q(λt); P_n,Q_n(t), relating the precessional motion in configuration space to the motion of the action angle variables in λ space. By means of the space ordered form (6.41) for the transition matrix T(λ) we can derive an explicit albeit formal expression for the canonical mapping, namely

$$\begin{pmatrix} a(\lambda) & -(b(\lambda^* t))^* \\ b(\lambda t) & (a(\lambda^*))^* \end{pmatrix} = \left(\exp\left(-i\lambda\int(\bar{S}(xt)\cdot\bar{\sigma} - \sigma^z)dx\right)\right)_+ \qquad (6.70)$$

together with (6.69a-c). Since the derivation of the Poisson bracket relations, in terms of which we have identified the canonical transformation, only involves differential statements, the construction of the action angle framework is a much simpler task than the explicit solution of the Gelfand-Levitan-Marchenko equation, discussed in Section 6.10.

The dynamical modes of the Heisenberg chain fall in two classes: Continuum modes characterised by the canonical variables $P(\lambda)$ and $Q(\lambda t)$ for $-\infty<\lambda<\infty$, and discrete modes specified by the canonical variables P_n and $Q_n(t)$, n = 1,2,..M. Since $P(\lambda)$ and P_n are constants of motion the continuum modes are characterised by the distribution $P(\lambda), -\infty<\lambda<\infty$, and the discrete modes by the complex numbers P_n, n = 1,2,..M, that is two real constants of motion for each mode. A given initial spin configuration $\bar{S}(xt)$ for the nonlinear evolution equation $d\bar{S}/dt = \bar{S}\times d^2\bar{S}/dx^2$ is, of course, equivalent to choosing a distribution $P(\lambda)$ for the continuum modes and a set of canonical momenta P_n for the discrete modes.

6.12 Energy - Momentum - Angular Momentum

For the purpose of investigating the physical properties of the continuum and discrete modes associated with the canonical action angle basis we now turn to the explicit construction of the Hamiltonian H, the total momentum Π, and the total angular momentum M^α in terms of the canonical variables $P(\lambda)$ and P_n.

From the time dependence of the canonical coordinates $Q(\lambda t)$ and $Q_n(t)$ given by Eqs. (6.65) and (6.68),

$$Q(\lambda t) = 4\lambda^2 t + \text{const.}$$

$$Q_n(t) = -\frac{4}{P_n^2}t + \text{const.}$$

in conjunction with the canonical equations of motion

$$\frac{dQ}{dt} = \{H, Q\}$$

$$\frac{dP}{dt} = \{H, P\}$$

and the Poisson bracket relations (6.64a-c) and (6.67a-c) we readily infer the form of the real Hamiltonian

$$H = \int d\lambda\, P(\lambda)(2\lambda)^2 + \sum_{n=1}^{M} 4\left(\frac{1}{p_n} + \frac{1}{p_n^*}\right) \tag{6.71}$$

In order to derive the action angle form of the total momentum we consider the Poisson bracket of Π with the transition matrix $T(\lambda)$. Using the property that the total momentum is the generator of translations in configuration space, i.e., from Eq. (6.10) $d\bar{S}/dx = -\{\Pi, \bar{S}\}$, we have

$$\{T(\lambda), \Pi\} = \int dx \sum_\alpha \frac{dT(\lambda)}{dS^\alpha(x)} \{S^\alpha(x), \Pi\} = \int dx \sum_\alpha \frac{dT(\lambda)}{dS^\alpha(x)} \frac{dS^\alpha(x)}{dx}$$

Inserting Eq.(6.61),

$$\frac{dT(\lambda)}{dS^\alpha(x)} = -i\lambda F^{-1}(x\lambda)\sigma^\alpha G(x\lambda),$$

performing a partial integration using the eigenvalue equation (6.34), $id\Psi/dx = \lambda S\Psi$, the idempotent property $S^2 = I$, and the boundary condition $S \to \sigma^z$ for $|x| \to \infty$, we obtain for λ real

$$\{T(\lambda), \Pi\} = -i\lambda\left(F^{-1}(\infty\lambda)\sigma^z G(\infty\lambda) - F^{-1}(-\infty\lambda)\sigma^z G(-\infty\lambda)\right)$$

Furthermore, introducing the transition matrix $T(\lambda)$ by Eq. (6.40), $G(x\lambda) = F(x\lambda)T(\lambda)$, and making use of the boundary conditions (6.37a-b), $F(x\lambda) \to \exp(-i\lambda\sigma^z x)$ for $x \to \infty$ and $G(x\lambda) \to \exp(-i\lambda\sigma^z x)$ for $x \to -\infty$, we arrive at

$$\{T(\lambda), \Pi\} = -i\lambda(\sigma^z T(\lambda) - T(\lambda)\sigma^z)$$

By analytic continuation into the Bargmann strip the only non vanishing Poisson brackets are

$$\{b(\lambda), \Pi\} = 2i\lambda b(\lambda), \quad -\infty < \lambda < \infty$$

$$\{b_n, \Pi\} = 2i\lambda_n b_n, \quad n=1,2,\dots M,$$

i.e., by the definitions (6.63b) and (6.66b)

$$\{Q(\lambda), \Pi\} = -2\lambda, \quad -\infty < \lambda < \infty$$

$$\{Q_n, \Pi\} = -\frac{2i}{P_n}, \quad n=1,2,\dots M$$

and the canonical Poisson brackets (6.64a-c) and (6.67a-c) imply that the real total momentum Π has the form

$$\Pi = \int d\lambda\, P(\lambda)\, 2\lambda + \sum_{n=1}^{M} 2i\left(\log P_n - \log P_n^* - i\pi\, \mathrm{sign}\,(\mathrm{Im}\, P_n)\right) \qquad (6.72)$$

Here the branch of the logarithm is defined such that the discrete contributions to H and Π both vanish for $\mathrm{Re}P_n \to 0$.

The total angular momentum M^α induces rotations in spin space. For the Poisson bracket of M^α with $T(\lambda)$ we obtain, using the definition (6.14a-c), $\vec{M} = \int dx(\vec{S}(x)-(0,0,1))$, and the spin algebra (6.2),

$$\{S^\alpha(x), S^\beta(y)\} = -\delta(x-y)\sum_\gamma \varepsilon^{\alpha\beta\gamma} S^\gamma(x),$$

$$\{T(\lambda), M^\alpha\} = \int dx \{T(\lambda), S^\alpha(x)\} = \sum_{\beta\gamma} \varepsilon^{\alpha\beta\gamma} \int dx \frac{dT(\lambda)}{dS^\beta(x)} S^\gamma(x).$$

Inserting Eq. (6.61) and using the identity

$$\frac{d}{dx}\left(F^{-1}(x\lambda)\sigma^\alpha G(x\lambda)\right) = -2\lambda \sum_{\beta\gamma} \varepsilon^{\alpha\beta\gamma} F^{-1}(x\lambda)\sigma^\beta G(x\lambda) S^\gamma(x)$$

we have for λ on the real axis

$$\{T(\lambda), M^\alpha\} = \frac{i}{2}\left(F^{-1}(\infty\lambda)\sigma^\alpha G(\infty\lambda) - F^{-1}(-\infty\lambda)\sigma^\alpha G(-\infty\lambda)\right).$$

In accordance with the boundary condition $S \to \sigma^z$ for $|x| \to \infty$ only $\{T(\lambda), M^z\}$ is well-defined. Introducing the transition matrix $T(\lambda)$ by Eq. (6.40) and using the boundary conditions (6.37a-b) for the Jost functions, we obtain

$$\{T(\lambda), M^z\} = \frac{i}{2}\left(\sigma^z T(\lambda) - T(\lambda)\sigma^z\right).$$

In the Bargmann strip the only non vanishing Poisson brackets are

$$\{b(\lambda), M^z\} = -ib(\lambda), \quad -\infty < \lambda < \infty$$

$$\{b_n, M^z\} = -ib_n, \quad n = 1, 2, \cdots M$$

i.e., by Eqs. (6.63b) and (6.66b)

$$\{Q(\lambda), M^z\} = 1, \quad -\infty < \lambda < \infty$$

$$\{Q_n, M^z\} = 1, \quad n = 1, 2, \cdots M$$

and we infer the following form for the real total angular mometum M^z:

$$-M^z = \int d\lambda\, P(\lambda) + \sum_{n=1} (P_n + P_n^*) \tag{6.73}$$

We notice that the total angular momentum of the dynamical modes is negative. This is evidently a consequence of our choice of ground state configuration $S = \sigma^z$, see the definition (6.14c).

6.13 The Spectrum of Solitons and Magnons

The explicit construction achieved in the previous section of the total energy, the total momentum, and the total angular momentum in terms of the canonical action angle variables allows for a simple interpretation of the spectrum of the classical Heisenberg chain in the long wavelength limit. Summarising, we found the expressions (6.71), (6.72), and (6.73),

$$H = \int d\lambda\, P(\lambda)\, 4\lambda^2 + \sum_{n=1}^{M} 4\left(\frac{1}{P_n} + \frac{1}{P_n^*}\right)$$

$$\Pi = \int d\lambda\, P(\lambda)\, 2\lambda + \sum_{n=1}^{M} 2i\left(\log\left(\frac{P_n}{P_n^*}\right) - i\pi\, \mathrm{sign}(\mathrm{Im}\, P_n)\right)$$

$$-M^z = \int d\lambda\, P(\lambda) + \sum_{n=1}^{M} (P_n + P_n^*).$$

The energy, momentum, and angular momentum are composed of two distinct contributions, a continuum part characterised by the real canonical momentum $P(\lambda)$, $-\infty < \lambda < \infty$, and a discrete part specified by the complex canonical momenta P_n, n = 1,2,...M.

In analogy with the treatment of the Sine Gordon equation[113] we interpret $P(\lambda)$ as the density of continuum modes on the λ axis. This characterisation is consistent with the action angle nature of the variables $P(\lambda)$ and $Q(\lambda)$. Hence, from the above expressions for H, Π , and M^z we conclude that the λth mode in units of $P(\lambda)$ has the energy, momentum, and angular momentum

$$\omega(\lambda) = 4\lambda^2 \quad \text{for} \quad -\infty < \lambda < \infty$$

$$\Pi(\lambda) = 2\lambda \quad \text{for} \quad -\infty < \lambda < \infty$$

$$m(\lambda) = -1 \quad \text{for} \quad -\infty < \lambda < \infty$$

The band of continuum modes can thus be characterised by the quadratic dispersion law

$$\omega(\lambda) = \Pi(\lambda)^2 \quad \text{for} \quad -\infty < \lambda < \infty, \tag{6.74}$$

and we identify them tentatively with the magnons or spin waves[98] treated in Sections 6.3 and 6.4. Notice, however, that the continuum modes considered here are subject to the fixed boundary condition $S(xt) \to \sigma^z$ for $|x| \to \infty$, unlike the spin waves discussed previously, which have a constant amplitude and are compatible only with periodic boundary conditions. The magnon band is completely characterised by the dispersion law (6.74) and the density $P(\lambda)$ in λ space. With our choice of ground state the angular momentum of the continuum modes is negative and has a magnitude equal to the integrated density $\int P(\lambda)d\lambda$.

The interpretation of the discrete contributions to H, Π, and M^z is more straightforward. The nth mode has the energy, momentum, and angular motentum

$$E_n = 4\left(\frac{1}{P_n} + \frac{1}{P_n^*}\right) \qquad \text{for } n = 1,2,..M$$

$$\Pi_n = 2i\left(\log\left(\frac{P_n}{P_n^*}\right) - i\pi\,\mathrm{sign}(\mathrm{Im}P_n)\right) \qquad \text{for } n = 1,2,..M$$

$$M_n = -(P_n + P_n^*). \qquad \text{for } n = 1,2,..M.$$

Introducing $P_n = A_n\exp(-i\theta_n/4)$, where since $\mathrm{Re}P_n > 0$ the phase θ_n is restricted to the interval $-2\pi < \theta_n < 2\pi$, and eliminating the amplitude A_n and the phase θ_n, we obtain the following dispersion law for the nth discrete mode

$$E_n = \frac{16}{|M_n|}\sin^2\left(\frac{\Pi_n}{4}\right), \quad -2\pi < \Pi_n < 2\pi, \quad n=1,2\cdots M. \tag{6.75}$$

This expression has exactly the same form as the dispersion law (6.8) for the permanent profile solitary wave discussed in detail in Section 6.4. The discrete modes found by the inverse scattering method can thus be identified with the solitary waves, and are according to modern terminology (see, for instance, refs. 89 and 90) called solitons. Unlike the magnons or continuum modes which are extended in space, the solitons are spatially localised objects with a width Γ_n given by Eq. 6.25, $\Gamma_n = 8/E_n$, i.e., inversely proportional to their energy. The solitons, furthermore, have an internal structure; they carry an angular momentum $-M_n$. In the limit of small momentum $|\Pi_n| \ll 1$, $E_n \simeq \Pi_n^2/|M_n|$, and we can associate an effective mass $|M_n|/2$ with the soliton. The "rest mass" $|M_n|/2$ is a function of the internal state of the soliton and is proportional to its angular momentum. For a more detailed discussion of the soliton or solitary wave we refer to Section 6.4.

The localised soliton modes and extended magnon modes thus diagonalise the Hamilton H and completely exhaust the spectrum of the classical

Heisenberg chain in the long wavelength limit. In Fig. 6.25 we have plotted the dispersion law (6.74) for the magnons and the dispersion law (6.75) for the solitons for different values of the angular momentum. Notice, however, that for the classical Heisenberg chain both magnons and solitons form bands in a plot of energy versus momentum (see Fig.6.8). We have shown the soliton band in Fig. 6.25.

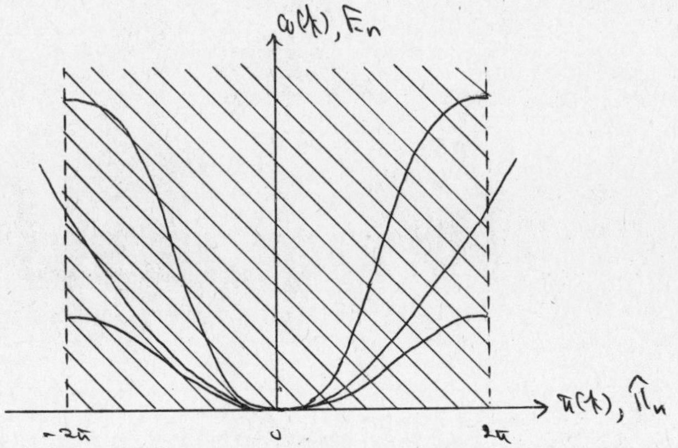

Fig. 6.26 The magnon dispersion law $\omega = \pi^2$ and the soliton dispersion law $E_n = 16\sin^2(\Pi_n/4)/|M_n|$ for two different values of M_n. The shaded area indicates the soliton band (arbitrary units).

Since the classical Heisenberg chain in "eigenvalue space" $\{\lambda\}$ is essentially a gas of non interacting magnon modes (radiation) and soliton modes (particles), the question of the stability of the solitons under collisions in configuration space, investigated numerically by Tjon and Wrigth[95], is immediately answered. Under soliton-soliton collision the constants of motion E, Π , and M for each individual soliton are preserved. Since the solitons are localised objects their shape long before and long after a collision is unaltered, as sketched in Fig. 6.26. In addition to the constant of motion E, Π , and M, related by the dispersion law (6.75), the dynamical state of a soliton is characterised by the center of mass x_0 and the phase ϕ_0, which do change under collision. The shifts Δx_0 and $\Delta\phi_0$ in the case of two soliton collision have been given by Takhatajan[102] on the basis of the Gelfand-Levitan-Marchenko equation (see also a magnon -soliton phase shift analysis by the present author[114]),

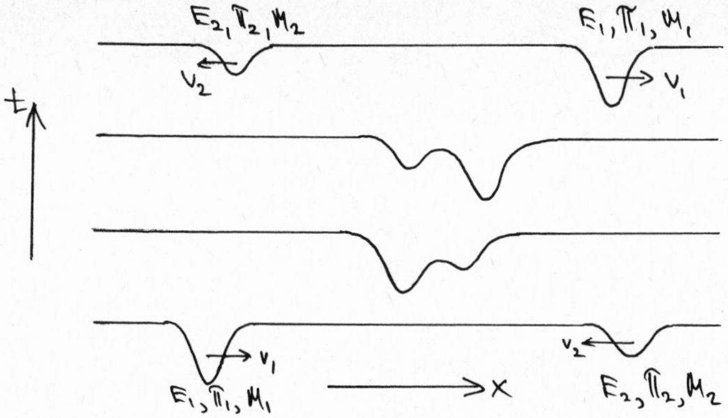

Fig. 6.27 Two soliton collision in configuartion space.

We emphasise that the identification of the continuum and discrete modes in canonical action angle space with the permanent profile spin and solitary wave solutions discussed in Section 6.4, is to some extent incomplete since we have not exhibited the explicit form in configuration space. However, as mentioned in Section 6.10, the Gelfand-Levitan-Marchenko equation reduces in the pure soliton case, i.e., a reflectionless potential, to a set of linear algebraic equations which are readily soluble at least in the case of few solitons, yielding the solitary wave form (see ref. 92). On the other hand, for the continuum modes, i.e., for a non vanishing reflection coefficient, the Gelfand-Levitan-Marchenko integral equation cannot be solved explicitly and the identification with spin waves remains tentative; see, however, ref. 114.

6.14 The Infinite Series of Constants of Motion

In Section 6.12 we derived the form of the three constants of motion H, Π , M^z associated with the global symmetry transformations: Time translation, space translation, and spin rotation. Since the long wavelength limit of the Heisenberg chain is a completely integrable Hamiltonian system it does, however, possess an infinite series of independent constants of motion in addition to H, Π , and M^z, which we now proceed to determine using the methods developed in refs. 113 and 115 in connection with the Sine Gordon and Korteweg-deVries equations.

According to Eq. (6.55a) the matrix element $a(\lambda)$ of the transition matrix $T(\lambda)$ is independent of time under the motion of $S(xt)$ and it follows in particular that the coefficients in a Laurent expansion of $\log a(\lambda)$ in powers of λ and $1/\lambda$ are constants of motion. By means of the spectral erepresentation (6.69a) we obtain

$$\mathrm{Im}(\log a(\lambda)) = -\sum_{K=1}^{\infty} \frac{1}{\lambda^K} A_K \quad \text{for} \quad |\lambda| \to \infty$$

$$\mathrm{Im}(\log a(\lambda)) = -\sum_{K=0}^{\infty} \lambda^K B_K \quad \text{for} \quad |\lambda| \to \infty,$$

where

$$A_K = \int d\lambda\, P(\lambda)\lambda^{1+K} + i\frac{(-1)^{K+1}}{K}\sum_{n=1}^{M} i^K((-1)^K p_n^{-K} - (p_n^*)^{-K})$$

$$B_0 = -\int d\lambda\, P(\lambda)\lambda + i\sum_{n=1}^{M}(\log p_n^* - \log p_n + i\pi\,\mathrm{sign}(\mathrm{Im}\,p_n))$$

$$B_K = -\int d\lambda\, P(\lambda)\lambda^{1-K} + i\frac{(-1)^{1+K}}{K}\sum_{n=1}^{M} i^{-K}((-1)^K p_n^K - (p_n^*)^K).$$

In order to determine the structure of the independent constants of motion A_k and B_k in configuration space, i.e., in terms of the spin density $S(xt)$, we establish a recursive procedure. From Eqs. (6.40) and (6.42) we have

$$G_{11}(x\lambda) = F_{11}(x\lambda)a(\lambda) + F_{12}(x\lambda)b(\lambda).$$

Introducing the density

$$\sigma(x\lambda) = \frac{d\log G_{11}(x\lambda)}{dx} + i\lambda$$

we obtain, since Eqs. (6.37a-b) $G_{11}(x\lambda) \to \exp(-i\lambda x)$ for $x \to -\infty$ and $G_{11}(x\lambda) \to a(\lambda)\exp(-i\lambda x)$ for $\lambda \to \infty$

$$\log a(\lambda) = \int \sigma(x\lambda)dx.$$

The density $\sigma(x\lambda)$ is now determined by considering the eigenvalue equation $i dG/dx = \lambda SG$ for the one-one component of the Jost function $G(x\lambda)$, i.e.,

$$i\frac{dG_{11}(x\lambda)}{dx} = \lambda(S^z(x)G_{11}(x\lambda) + S^-(x)G_{21}(x\lambda)).$$

Introducing $\Phi(x\lambda) = G_{21}(x\lambda)/G_{11}(x\lambda)$ we have

$$\sigma(x\lambda) = -i\lambda(S^z(x) - 1 + S^-(x)\Phi(x\lambda)),$$

i.e.,

$$\log a(\lambda) = -i\lambda\int(S^z(x) - 1 + S^-(x)\Phi(x\lambda))dx,$$

where the auxiliary function $\Phi(x\lambda)$ satisfies the non linear generalised Ricatti equation[116]

$$i\frac{d\Phi(x\lambda)}{dx} + \lambda(S^-(x)\Phi(x\lambda)^2 + 2S^z(x)\Phi(x\lambda) - S^+(x)) = 0$$

with the boundary condition $\Phi(-\infty, \lambda) = 0$. It is now easy to derive expansions in powers of λ and $1/\lambda$. Inserting $\Phi(x\lambda) = \sum_{n=1}^{\infty} \lambda^n f_n(x)$ and $\Phi(x\lambda) = \sum_{n=0}^{\infty} \lambda^{-n} g_n(x)$ in the above Riccatti equation and comparing terms we obtain the recursion formulae

$$i\frac{df_n}{dx} + S^-\sum_{p=0}^{n-1} f_p f_{n-p-1} + 2S^z f_{n-1} - S^+\delta_{n1} = 0, \quad f_0 = 0$$

$$i\frac{dg_n}{dx} + S^-\sum_{p=0}^{n+1} g_p g_{n-p+1} + 2S^z g_{n+1} = 0, \quad g_0 = \frac{1-S^z}{S^-}.$$

By straightforward iteration we find

$$\Phi(x\lambda) = -i\lambda\int_{-\infty}^{x} S^+(y)dy - \lambda^2\int_{-\infty}^{x} dy\, S^z(y)\int_{-\infty}^{y} dz\, S^+(z) + \dots \qquad \text{for } \lambda \to 0$$

$$\Phi(x\lambda) = \frac{1-S^z(x)}{S^-(x)} - \frac{i}{2\lambda}\frac{d}{dx}\left(\frac{1-S^z(x)}{S^-(x)}\right) + \dots \qquad \text{for } \lambda \to \infty$$

which by substitution in the expressions for $\log a(\lambda)$ and $\mathrm{Im}(\log a(\lambda))$ yield the conserved densities $a_n(x)$ and $b_n(x)$ associated with the constants of motion A_n and B_n, i.e., $A_n = \int dx a_n(x)$ and $B_n = \int dx b_n(x)$,

$\vdots$

$$a_1(x) = \frac{1}{8}\left(\left(\frac{dS^x}{dx}\right)^2 + \left(\frac{dS^y}{dx}\right)^2 + \left(\frac{dS^z}{dx}\right)^2\right) \quad \left(= \frac{1}{4}\varepsilon(x)\right)$$

$$b_0(x) = -\frac{1}{4}\,\frac{S^y(x)\frac{dS^x(x)}{dx} - S^x(x)\frac{dS^y(x)}{dx}}{1 + S^z(x)} \quad \left(= -\frac{1}{2}\pi(x)\right)$$

$$b_1(x) = S^z(x) - 1 \quad (= m(x))$$

$$b_2(x) = S^x(x)\int_{-\infty}^{x} S^y(y)\,dy - S^y(x)\int_{-\infty}^{x} S^x(y)\,dy$$

$$b_3(x) = -S^x(x)\int_{-\infty}^{x} dy\, S^z(y)\int_{-\infty}^{y} dz\, S^x(z) - S^y(x)\int_{-\infty}^{x} dy\, S^z(y)\int_{-\infty}^{y} dz\, S^y(z)$$

$$\vdots$$

As expected the energy, momentum, and angular momentum densities are included in the infinite series of conserved densities, i.e., $\mathcal{E}(x) = 4a_1(x)$, $\Pi(x) = -2b_0(x)$, and $m(x) = b_1(x)$. We notice that the densities fall in two classes. The ones associated with the expansion of $\log a(\lambda)$ in powers of λ, $b_n(x)$, $n = 1,2,..$, include the angular momentum density $b_1(x)$, and are for $n > 2$ non local functions of the spin density $S(x)$. They can, incidentally, also be derived by expanding the formal space ordered expression (6.70) for the non linear canonical transformation in powers of λ. On the other hand, the conserved densities $b_0(x)$ and $a_n(x)$, $n = 1,2,..$, arising from expanding $\log a(\lambda)$ in powers of $1/\lambda$, include the energy and momentum densities $4a_1(x)$ and $-2b_0(x)$, and are local functions of the spin density $S(x)$ and its derivatives $dS(x)/dx$, etc. The local densities $a_1 \simeq \mathcal{E}$, $b_0 \simeq \Pi$, and $b_1 \simeq m$ are related to the global symmetries: Time translation, space translation, and spin rotation. The question of whether the other conserved densities are associated with underlying local symmetries (gauge symmetries!) is a fascinating one, but so far essentially unexplored (see, however, ref. 89).

6.15 Summary and Conclusion

In the preceeding sections we have carried out a detailed analysis of certain aspects of the dynamical behaviour of the classical isotropic Heisenberg chain in the long wavelength limit. By means of the "russian version" of the inverse scattering techniques we amplified the work of Takhtajan[102] and exhibited in particular the canonical action angle and localise soliton modes. We remark, however, that unlike a gas of point-like particles, the solitons have a finite size and an internal structure.

Their dynamical behaviour in configuarion space is therefore non trivial since they under collision temporarily change their shape and furthermore suffer permanent shifts of positions and phases, while still preserving their constants of motion. A detailed understanding of the dynamics of "non interacting" magnons and solitons in configuration space requires, however, an anlysis of the Gelfand -Levitan-Marchenko integral equation, so far only carried out in the pure soliton case[102], but clearly representing an interesting field of further studies (see ref. 114).

We remark in passing that the canonical action angle representation allows for a semi classical quantisation according to standard rules[8] (see also ref. 117) by simply replacing the Poisson brakckets by commutators. Owing to the uncertainty principle the soliton mode thus becomes delocalised and both solitons and magnons appear as elementary excitations on an equal footing. The quantum Heisenberg chain in the semi classical limit, i.e., the limit of large S since $S\hbar \rightarrow 1$ for $\hbar \rightarrow 0$, thus consists of two kinds of non interacting bosons: Spin one magnons with a quadratic dispersion law $E = p^2$ and solitons with an arbitrary integer spin ν and a dispersion law $E = 16 \sin^2(p/4)/\nu$, $\nu = 1,2,\ldots$ ($\hbar = 1$). We notice the interesting feature that the classical soliton band in Fig. 6.25 under quantisation breaks up into separate dispersion laws labelled by the spin quantum number ν. Furthermore, in the low momentum limit $p \ll 1$, i.e., in the long wavelength limit, the effective soliton mass m/2 is quantised in half integer units. Since, as is well-known, the classical canonical transformation does not correspond to a uniquequantum mechanical unitary mapping, the problem of calculating quantum corrections to the semi classical limit is a subtle one (see refs. 118-120).

The action angle representation does also provide the natural starting point for constructing the statistical mechanics of the Heisenberg chain from "first principles", as well as understanding the influence of perturbations such as anisotropy, impurities, finite lattice distance effects, etc (see ref. 88 and 121).

Finally, we mention that quite recently it has been shown[122] that the continuous Heisenberg chain is equivalent to the non linear Schrödinger equation[105]; this correspondence corroborates the integrability demonstrated in ref. 111.

LIST OF REFERENCES

1. J.B. Torrance, Jr., and M. Tinkham, Phys. Rev. 187, 587 (1969)
2. J.B. Torrance, Jr., and M. Tinkham, Phys. Rev. 187, 595 (1969)
3. D.F. Nicoli and M. Tinkham, Phys. Rev. B9, 3126 (1974)
4. H.C. Fogedby, Phys. Rev. B5, 1941 (1972)
5. H.C. Fogedby and H. Højgaard Jensen, Phys. Rev. B6, 3444 (1972)
6. H.C. Fogedby, Phys. Rev. B8, 2200 (1973)
7. H.C. Fogedby, Phys. Rev. B10, 4000 (1974)
8. L.D. Landau and E.M. Lifschitz, Quantum Mechancis (Pergamon, London 1958)
9. E. Ising, Z. Phys. 31, 253 (1925)
10. H.A. Bethe, Z. Phys. 71, 205 (1931)
11. F. Bloch, Z. Phys. 61, 206 (1930)
12. C. Kittel, Introduction to Solid State Physics (Wiley, London 1971)
13. R. Orbach, Phys. Rev. 112, 309 (1958)
14. M. Wortis, Phys.Rev. 132, 85 (1963)
15. J. Hanus, Phys. Rev. Lett. 11, 336 (1963)
16. J.E. van Himbergen and J.A. Tjon, Physica 76, 503 (1974)
17. R. Courant and D. Hilbert, Methods of Mathematical Physics (Interscience, New York 1966)
18. E. Janhke and F. Emde, Tables of Functions with Formulae and Curves (Dover, New York, 1945)
19. H. Højgaard Jensen (private communication)
20. P.C. Martin, Houches Lectures, 1967; edited by C.N. DeWitt and R. Balian (Gordon and Breach, New York 1967)
21. D. Forster, Hydrodynamic Fluctuations. Broken Symmetry and Correlation Functions (Benjamin, Reading, Mass. USA 1975)
22. J.B. Torrance, Jr., Ph.D. thesis (Harvard University, 1968) (unpublished); Harvard University, Division of Engineering and Applied Physics, Technical Report No. 1, 1969 (unpublished)

23. H.C. Fogedby, unpublished

24. A.A. Abrikosov, L.P. Gorkov, and I.E. Dzyaloshinskii, Methods of Quantum Field Theory in Statistical Physics (Pergamon, Oxford 1965)

25. N.N. Bogoliubov and D.V. Shirkov, Introduction to the Theory of Quantised Fields (Interscience, New York 1959)

26. H.C. Fogedby, J.Phys. C11, 2801 (1978)

27. S.A. Pikin and V.M. Tsukernik, Sov. Phys.-JETP 23, 914 (1965)

28. P. Pfeuty, Ann. Phys. (N.Y.) 57, 79 (1970)

29. T.N. Tommet and D.L. Huber, Phys. Rev. B11, 450 (1975)

30. P. Jordan and E. Wigner, Z. Phys. 47, 631 (1928)

31. E. Lieb, T. Schultz, and D. Mattis, Ann.Phys. (N.Y.) 16, 407 (1961)

32. R. Jullien, P. Pfeuty, J.N. Fields, and S. Doniach, Phys. Rev. B18, 3568 (1978)

33. N.N. Lebedev, Special Functions and their Applications (Dover, New York 1972)

34. B. Southern and F.D.M. Haldane (private communication)

35. G. Toulouse and P. Pfeuty, Introduction au Groupe de Renormalisation et a ses Applications (Presses Universitaires, Grenoble 1975)

36. S-k. Ma, Modern Theory of Critical Phenomena (Benjamin, Reading, Mass. USA 1976)

37. H.C. Fogedby, J. Phys. C11, 4767 (1978)

38. H.C. Fogedby, J. Phys. C9 , 3757 (1976)

39. H.C. Fogedby, J. Phys. C10, 2869 (1977)

40. H.C. Fogedby, J. Phys. C11, 969 (1978)

41. B.M McCoy, Phys. Rev. 173, 531 (1968)

42. Th. Niemeyer, Physica 36, 377 (1967)

43. B.M. McCoy, E. Barouch, and D.B. Abraham, Phys. Rev. A4, 2331 (1971)

44. S.Katsura, Phys. Rev. 127, 1508 (1962)

45. B. Sutherland, J. Math. Phys. 11, 3183 (1970)

46. R.J. Baxter, Phys. Rev. Lett. 26, 832 (1971); Ann. Phys. (N.Y.) 70, 193 (1972)

47. J.D. Johnson, J. Krinsky, and B.M. McCoy, Phys. Rev. A8, 2526 (1973)

48. B.M. McCoy and T.T. Wu, The Two-Dimensional Ising Model (Harvard University Press, Cambrdige, Mass. 1973)

49. A. Luther and I. Peschel, Phys. Rev. B12, 3908 (1975)

50. A. Luther and I. Peschel, Phys. Rev. B9, 2911 (1974)

51. J.M Luttinger, J. Math. 4, 1154 (1963)

52. S. Tomonaga, Prog. Theor. Phys. 5, 544 (1950)

53. P.A. Wolff, Phys. Rev. 124, 1030 (1961)

54. P.C. Martin and Schwinger, Phys. Rev. 115, 1342 (1959)

55. L.P. Kadanoff and G. Baym, Quantum Statistical Mechanics (Benjamin, New York 1962)

56. M. Wortis, Ph.D. thesis (Harvard University, 1963) (unpublished)

57. T. Holstein and H. Primakoff, Phys. Rev. 58, 1098 (1940)

58. D.C. Mattis and E.H. Lieb, J. Math. Phys. 6, 304 (1965)

59. A. Theumann, J. Math. Phys. 8, 2460 (1967)

60. C.B. Dover, Ann. Phys. (N.Y.) 50, 500 (1968)

61. R.L. Statonovich, Sov. Phys. Dokl. 2, 416 (1957)

62. J. Hubbard, Phys. Rev. Lett. 3, 77 (1959)

63. J. Des Cloiseaux and J.J. Pearson, Phys. Rev. 128, 2131 (1962)

64. M. Steiner, J. Villain, and C.G. Windsor, Adv. Phys. 25, 87 (1976)

65. H.C. Fogedby and A.P. Young, J. Phys. C11, 527 (1978)

66. L.D. Landau and E.M. Lifshitz, Statistical Physics (Pergamon, London 1959)

67. L.P. Kadanoff and P.C. Martin, Ann. Phys. (N.Y.) 24, 419 (1963)

68. L.D. Landau and E.M. Lifshitz, Mechanics (Pergamon, London 1960)

69. B.I. Haperin, P.C. Hohenberg, and S-k- Ma, Phys. Rev. B10, 139 (1974)

70. P.C. Hohenberg and B.I. Halperin, Rev. Mod. Phys. 49, 435 (1977)

71. L.D. Landau and E.M. Lifshitz, Fluid Mechanics (Pergamon, London 1959)

72. R.L. Stratonovich, Topics in the Theory of Random Noise, Vol. 1 (Gordon and Breach, New York 1963)

73. R. Graham and H. Haken, Z. . 243, 289 (1971)

74. R. Graham and H. Haken, Z. Phys. 243, 141 (1971)

75. S-k. Ma and G.F. Mazenko, Phys. Rev. B11, 4077 (1975)

76. L. Sasvari and P. Szepfalusy, Physica 87A, 1 (1977)

77. Y. Pomeau and P. Resibois, Phys. Rev. C19, 64 (1975)

78. M. Månson, J. Phys. F7, 4073 (1974)

79. P. Borckmans, G. Dewel and D. Walgraef, Physica 88A, 261 (1977)

80. D. Forster, D.R. Nelson, and M.J. Stephen,Phys. Rev. Lett. 36, 867 (1976)

81. D. Forster, D.R. Nelson, and M.J. Stephen, Phys. Rev. A16, 732 (1977)

82. A.M. Polyakov, Phys. Lett. B59, 79 (1975)

83. A.A. Migdal, Sov. Phys.-JETP 42, 743 (1976)

84. E. Brézin and J. Zinn-Justin, Phys. Rev. Lett. 36, 691 (1976)

85. E. Brézin and J. Zinn.Justin, Phys. Rev. B14, 3110 (1976)

86. D.S. Amit and S-k- Ma (unpublished)

87. D.R. Nelson and D.S. Fisher, Phys. Rev. B16, 4945 (1977)

88. H.C. Fogedby, J. Phys. A13, 1467 (1980)

89. A.C. Scott, F.Y.F. Chu, and D.W. McLaughlin, Proceed. IEEE. 61, 1443 (1973)

90. Solitons and Condensed Matter Physics, edited by A.R. Bishóp and T. Schneider (Springer, Berlin 1979)

91. L.D. Landau and E.M. Lifshitz, Phys. Z. Sowjet 8, 153 (1935)

92. W. Döring, Z. Phys. 124, 501 (1947)

93. N.D. Mermin, J. Math. Phys. 8, 1061 (1967)

94. N.D. Mermin, Phys. Rev. 134, A112 (1964)

95. J. Tjon and J. Wright, Phys. Rev. B15, 3470 (1977)

96. E.T. Whittaker and G.N. Watson, A Course of Modern Analysis (University Press, Cambridge 1962)

97. H. Jeffreys and B. Jeffereys, Methods of Mathematical Physics (University Press, Cambridge 1972)

98. M. Lakshmanan, Th. W. Ruijgrok, and C.J. Thompson, Physica 84A, 577 (1976)

99. K. Nakamura and T. Sasada, Phys. Lett. 48A, 321 (1974)

100. G.B. Whitham, Linear and Non Linear Waves, (Wiley & Sons, New York 1974)

101. L.D. Landau and E.M. Lifshitz, The Classical Theory of Fields, (Pergamon, London 1962)

102. L.A. Takhtajan, Phys. Lett. 64A, 235 (1977)

103. C.S. Gardner, J.M. Green, M.D. Kruskal, and R.M. Miura, Phys. Rev. Lett. 19, 1095 (1967)

104. P.D. Lax, Commun. Pure Appl. Math. 21, 467 (1968)

105. V.E.Zakharov and A.B. Shabat, Sov. Phys.-JETP 34, 62 (1972)

106. M.J. Ablowitz, D.J. Kaup, A.C. Newell, and H. Segur, Phys. Rev. Lett. 30, (1973)

107. L.D. Faddeev, J. Math. Phys. 4, 72 (1963)

108. E. Goursat, A Course in Mathematical Analysis (Dover, New York 1964)

109. G.L. Lamb Jr., Phys. Rev. Lett. 37, 235 (1976)

110. G.L. Lamb Jr., J. Math. Phys. 18, 1654 (1977)

111. M. Lakshmanan, Phys. Lett. 61A, 53 (1977)

112. V.E. Zakharov and S.V. Manakov, Theor. Math. Phys. 19, 551 (1975)

113. L.A. Takhtajan and L.D. Faddeev, Theor. Math. Phys. 21, 1046 (1975)

114. H.C. Fogedby, Physica Scripta (to be published)

115. V.E. Zakharov and L.D. Faddeev, Funct. Anal. Appl. 5, 280 (1972)

116. E.L. Ince, Ordinary Differential Equations (Dover, New York 1956)

117. V.E. Korenpin and L.D. Faddeev, Theor. Math. Phys. 25, 1039 (1976)

118. H.C. Fogedby, J. Phys. C13, L195 (1980)

119. A. Jevicki and M. Papanicolaou, Ann. Phys. (N.Y.) 120, 107 (1979)

120. P.P. Kulish and E.K. Sklyanin, Phys.Lett. 70A, 461 (1979).

121. K.A. Long and A.P. Bishop, J. Phys. A.12 , 1325 (1979)

122. V.E. Zakharov and L.A. Takhtajan, Theor. Mat. Phys. 38, 17 (1979)

Condensed Matter

Zeitschrift für Physik B

Subscription Information:
1980. Volumes 36–39 (4 issues each):
Sample copy upon request.

All countries (except North America):
1980. DM 784,–, plus postage and handling.
Send your order or request
to your bookseller or directly to:
Springer-Verlag, Wissenschaftliche
Information Zeitschriften,
Postfach 105 280, D-6900 Heidelberg, FRG

North America:
1980. US $ 456.00, including postage and handling. Subscriptions are entered with prepayment only. Send your order or request to your bookseller or to:
Springer-Verlag New York Inc.,
175 Fifth Avenue, New York,
NY 10010, USA

Europhysics Journal

ISSN 0340-224X Title No. 257

Zeitschrift für Physik appears in three parts:

A: Atoms and Nuclei
B: Condensed Matter
C: Particles and Fields

Each part may be ordered separately.
Coordinating editor for Zeitschrift für Physik, Parts A, B and C: O. Haxel, Heidelberg

**ZEITSCHRIFT FÜR PHYSIK B
CONDENSED MATTER**

Physics of Condensed Matter
Physical properties of crystalline, disordered and amorphous solids
Classical and quantum-fluids
Topics of molecular physics related to the physics of condensed matter

General Physics
Quantum optics
Statistical physics, nonequilibrium and cooperative phenomena

Springer-Verlag
Berlin
Heidelberg
New York

Selected Issues from

Lecture Notes in Mathematics

Vol. 712: Equations Différentielles et Systèmes de Pfaff dans le Champ Complexe. Edité par R. Gérard et J.-P. Ramis. V, 364 pages. 1979.

Vol. 716: M. A. Scheunert, The Theory of Lie Superalgebras. X, 271 pages. 1979.

Vol. 720: E. Dubinsky, The Structure of Nuclear Fréchet Spaces. V, 187 pages. 1979.

Vol. 724: D. Griffeath, Additive and Cancellative Interacting Particle Systems. V, 108 pages. 1979.

Vol. 725: Algèbres d'Opérateurs. Proceedings, 1978. Edité par P. de la Harpe. VII, 309 pages. 1979.

Vol. 726: Y.-C. Wong, Schwartz Spaces, Nuclear Spaces and Tensor Products. VI, 418 pages. 1979.

Vol. 727: Y. Saito, Spectral Representations for Schrödinger Operators With Long-Range Potentials. V, 149 pages. 1979.

Vol. 728: Non-Commutative Harmonic Analysis. Proceedings, 1978. Edited by J. Carmona and M. Vergne. V, 244 pages. 1979.

Vol. 729: Ergodic Theory. Proceedings 1978. Edited by M. Denker and K. Jacobs. XII, 209 pages. 1979.

Vol. 730: Functional Differential Equations and Approximation of Fixed Points. Proceedings, 1978. Edited by H.-O. Peitgen and H.-O. Walther. XV, 503 pages. 1979.

Vol. 731: Y. Nakagami and M. Takesaki, Duality for Crossed Products of von Neumann Algebras. IX, 139 pages. 1979.

Vol. 733: F. Bloom, Modern Differential Geometric Techniques in the Theory of Continuous Distributions of Dislocations. XII, 206 pages. 1979.

Vol. 735: B. Aupetit, Propriétés Spectrales des Algèbres de Banach. XII, 192 pages. 1979.

Vol. 738: P. E. Conner, Differentiable Periodic Maps. 2nd edition, IV, 181 pages. 1979.

Vol. 742: K. Clancey, Seminormal Operators. VII, 125 pages. 1979.

Vol. 755: Global Analysis. Proceedings, 1978. Edited by M. Grmela and J. E. Marsden. VII, 377 pages. 1979.

Vol. 756: H. O. Cordes, Elliptic Pseudo-Differential Operators – An Abstract Theory. IX, 331 pages. 1979.

Vol. 760: H.-O. Georgii, Canonical Gibbs Measures. VIII, 190 pages. 1979.

Vol. 762: D. H. Sattinger, Group Theoretic Methods in Bifurcation Theory. V, 241 pages. 1979.

Vol. 765: Padé Approximation and its Applications. Proceedings, 1979. Edited by L. Wuytack. VI, 392 pages. 1979.

Vol. 766: T. tom Dieck, Transformation Groups and Representation Theory. VIII, 309 pages. 1979.

Vol. 771: Approximation Methods for Navier-Stokes Problems. Proceedings, 1979. Edited by R. Rautmann. XVI, 581 pages. 1980.

Vol. 773: Numerical Analysis. Proceedings, 1979. Edited by G. A. Watson. X, 184 pages. 1980.

Vol. 775: Geometric Methods in Mathematical Physics. Proceedings, 1979. Edited by G. Kaiser and J. E. Marsden. VII, 257 pages. 1980.

Vol. 779: Euclidean Harmonic Analysis. Proceedings, 1979. Edited by J. J. Benedetto. III, 177 pages. 1980.

Vol. 780: L. Schwartz, Semi-Martingales sur des Variétés, et Martingales Conformes sur des Variétés Analytiques Complexes. XV, 132 pages. 1980.

Vol. 782: Bifurcation and Nonlinear Eigenvalue Problems. Proceedings, 1978. Edited by C. Bardos, J. M. Lasry and M. Schatzman VIII, 296 pages. 1980.

Vol. 783: A. Dinghas, Wertverteilung meromorpher Funktionen in ein- und mehrfach zusammenhängenden Gebieten. Edited by R. Nevanlinna and C. Andreian Cazacu. XIII, 145 pages. 1980.

Vol. 786: I. J. Maddox, Infinite Matrices of Operators. V, 122 pages 1980.

Vol. 787: Potential Theory, Copenhagen 1979. Proceedings, 1979. Edited by C. Berg, G. Forst and B. Fuglede. VIII, 319 pages. 1980.

Vol. 791: K. W. Bauer and S. Ruscheweyh, Differential Operators for Partial Differential Equations and Function Theoretic Applications. V, 258 pages. 1980.

Vol. 792: Geometry and Differential Geometry. Proceedings, 1979. Edited by R. Artzy and I. Vaisman. VI, 443 pages. 1980.

Vol. 793: J. Renault, A Groupoid Approach to C*-Algebras. III, 160 pages. 1980.

Vol. 798: Analytic Functions, Kozubnik 1979. Proceedings. Edited by J. Ławrynowicz. X, 476 pages. 1980.

Vol. 799: Functional Differential Equations and Bifurcation. Proceedings 1979. Edited by A. F. Izé. XXII, 409 pages. 1980.

Vol. 801: K. Floret, Weakly Compact Sets. VII, 123 pages. 1980.

Vol. 802: J. Bair, R. Fourneau, Etude Géometrique des Espaces Vectoriels II. VII, 283 pages. 1980.

Vol. 804: M. Matsuda, First Order Algebraic Differential Equations. VII, 111 pages. 1980.

Vol. 805: O. Kowalski, Generalized Symmetric Spaces. XII, 187 pages. 1980.

Vol. 807: Fonctions de Plusieurs Variables Complexes IV. Proceedings, 1979. Edited by F. Norguet. IX, 198 pages. 1980.

Vol. 810: Geometrical Approaches to Differential Equations. Proceedings 1979. Edited by R. Martini. VII, 339 pages. 1980.

Vol. 816: L. Stoica, Local Operators and Markov Processes. VIII, 104 pages. 1980.

Vol. 819: Global Theory of Dynamical Systems. Proceedings, 1979 Edited by Z. Nitecki and C. Robinson. IX, 499 pages. 1980.